VOLCANO

The Earth series traces the historical significance and cultural history of natural phenomena. Written by experts who are passionate about their subject, titles in the series bring together science, art, literature, mythology, religion and popular culture, exploring and explaining the planet we inhabit in new and exciting ways.

Series editor: Daniel Allen

In the same series

Waterfall Brian J. Hudson

Volcano

James Hamilton

REAKTION BOOKS

For my mother, who loves volcanoes.

Published by
Reaktion Books Ltd
33 Great Sutton Street
London EC1V 0DX, UK
www.reaktionbooks.co.uk

First published 2012

Printed and bound in China by Eurasia

British Library Cataloguing in Publication Data
Hamilton, James, 1948–
Volcano : nature and culture. – (The Earth)
1. Volcanoes in art. 2. Volcanic eruptions – Social aspects.
3. Volcanoes – Mythology.
I. Title II. Series
704.9'4955121-dc23

ISBN 978 1 86189 917 0

CONTENTS

Preface

> 'Maybe it is not the destructiveness of the volcano that pleases most, though everyone loves a conflagration, but its defiance of the law of gravity to which every inorganic mass is subject . . . Perhaps we attend to a volcano for its elevation, like ballet. How high the molten rocks soar, how far above the mushrooming cloud. The thrill is that the mountain blows itself up, even if it must then like the dancer return to earth; even if it does not simply descend – it falls, it falls on us.'
>
> Susan Sontag, *The Volcano Lover*

The eruption in April 2010 beneath the Icelandic glacier Eyjafjallajökull was a salutary reminder of the power of volcanoes. This is a small volcano, but nevertheless its cloud of smoke and ash, directed across Britain and the Continent by an inconvenient prevailing wind, drove a crucial British General Election out of the main headlines for nearly a week, and threw world air traffic into temporary disarray.

We live on the same cooling planet that experienced eruptions of Santorini around 1620 BC, Vesuvius in AD 79 and 1631, Hekla in 1766–8, Tambora in 1815 and Krakatoa in 1883, all to devastating effect. These eruptions are part of a continual geological process, and what Eyjafjallajökull brought clearly to mind is that an ill-timed volcanic eruption of sufficient scale and unfortunate location will have such an immediate effect on the earth's climate and economy as to make global warming seem slow and dull by comparison.

At any time there's a volcano erupting somehow, somewhere, with clouds of ash, and the accompanying thunder and lightning *obbligato*. What the Icelandic eruption did was to bring home some new understanding of the power of volcanoes, and of the fragility of our planet.

This book, which has grown out of the exhibition 'Volcano: From Turner to Warhol' held in 2010 at Compton Verney, Warwickshire, explores artists' and writers' perception of volcanoes and its change over time.

I 'The Whole Sea Boiled and Blazed'

When we follow humanity in its earliest forms, creeping tentatively out from the African Rift Valley or, later, from what is now Australia, and moving gradually and ignorantly across the oceans by means of Continental Drift, we are traversing a period of two and a half million years or more. Adding on another ten or twelve thousand years, a blink of an eye by comparison, we can watch settlement in the Middle East, in the Americas, continental Europe and Southeast Asia. Here we arrive at the beginnings of ritual, the making of tools, and the first groupings of people cooperating to develop agriculture, the beginning of language and narrative, and the beginning too of accounting and writing. Across all these expanses of time volcanoes blasted from much the same weak points in the earth's surface as they had blasted for aeons before. It's all the same to them. They carry on in much the same way today. Many weak points have cooled down and dried up, but the map of the earth's cracks, now, bears a clear relationship to the map of the crust before Continental Drift, and is a direct product of the earth's cooling movements.

Volcanoes were at work long before any form of humanity began to populate the planet and take notice. Thus to human history, though both localized and scattered, they are a given, constant presence. As the most violent terrestrial outrage that the planet can offer and humanity can witness, volcanic action may thus be the source of the first faint distant tracings of narrative on human memory.

Çatal Hüyük , the world's first town, which developed 7500 to 5700 BC in a fertile volcanic landscape.

Such tracings come down to us intermittently and partially. We only have such artefacts as have been discovered, and such myths as have been recorded. When the volcano Hasan Dağ in the Karapinar field in Anatolia, western Turkey, erupted in around 6200 BC, it caused sufficient disturbance and alarm in Çatal Hüyük, 90 miles (145 km) away, to seep into the local consciousness of this very large and long-established settlement, and become the subject of a wall painting of the volcano erupting over tightly packed houses.[1] Çatal Hüyük, which was first excavated in the late 1950s and into the following decade, was by the standards of this early period a sophisticated place, the world's first town of up to about 10,000 people. The inhabitants, who lived in close-packed mud huts with white plastered interiors and rudimentary furniture, created a rural economy through farming the surrounding Anatolian plain. While the wall painting may have been a direct reaction, however long delayed it might have been, to the eruption, the fertility of the soil and thus the people's living was the product of the eruptive local volcano. This society had been settled long enough for it to develop its own manufacturing infrastructure, from pottery to plough-making, rising in sophistication to the programmatic creation of wall paintings and the making of blades from obsidian, a volcanic product.[2]

If the residents of Çatal Hüyük had developed a narrative tradition in which the volcano played a part, it has not come down to us. It is not for another 4,000 years that myth begins to shade into history when the volcano Santorini, now named Thera, the island in the Aegean Sea more or less equidistant between mainland Greece and Turkey, exploded around 1620 BC in what was the greatest event of natural destruction in recorded human history. Nearby Akrotiri was buried under lava and ash, while the tsunami that the eruption generated grew into a ten-metre high wave. This, after an unbroken run to Crete, cast itself on Knossos. The immediate destruction was one of the factors that precipitated the downfall of the Minoan civilization and, radiating in all directions, the tsunami brought havoc to the Aegean and its littoral. The Santorini eruption is one of the suggested causes of the disappearance of Atlantis – if, indeed, Atlantis ever existed.

A group of volcanoes whose perturbations also left their traces on classical myth are the Lipari islands, a volcanic chain north of Sicily. The southernmost of these, Vulcano, which passed through a long eruptive period around 400 BC, was explained in Greek (and Roman) myth as the furnace and forge of Hephaistos, the god of fire, or his Roman equivalent, Vulcan. When the mountain erupted, as it did regularly in classical times, it was thought to signal that Hephaistos was at work. The Lipari islands were both deadly and convenient, as Demeter (Roman: Ceres), the goddess of the fertility of the earth, used them, and their near

Jonas Umbach, *The Elements: Fire*, *c.* 1645–1700, etching. Vulcan is pictured among the clouds; his forge is on the left and cannons on the right.

neighbours, Vesuvius, Ischia and Etna, as torches to light the way for the Sirens in their search for her lost daughter Persephone. The 400 BC eruption of Vulcano may have been the source of Thucydides' observation in *The Peloponnesian War* that 'at night great flames are seen rising up, and in day-time the place is under a cloud of smoke'.[3] Mount Etna on Sicily, part of the same volcanic system, was also said to be the workshop of Hephaistos/Vulcan, here accompanied by one-eyed giants, the Cyclopes. These were the same bad-tempered oversized monsters who were so inhospitable to Odysseus (Ulysses) when he and his companions landed on their island in Homer's *Odyssey*.[4] In a connected myth, it was believed that the repulsive multi-headed giant Typhon, offspring of Gaia, the primal goddess of the earth, and Tartarus, the god of the wind, was imprisoned by Zeus, where 'he's lying just beside the straits of the sea, trapped beneath the roots of Mount Etna'.[5] Tossing and turning still, he causes the ructions and eruptions that are a regular and continuing feature of Etna. Taking the myth of the Cyclopes yet deeper, the single, circular eye of this beast is reflected in the single, circular form of the crater: thus the volcano becomes the creature within. Does Homer's epic tale, itself a late compilation of centuries of oral

Bernard Picart, *Enceladus buried beneath Mount Etna*, 1731, engraving.

ENCELADE PRÉCIPITÉ SOUS LE MONT ETHNA.
Enceladus burried under Mount Æthna.

Enceladus unter dem Berg Æthna bedeckt.
Enceladus onder den Berg Ethna bedekt.

Eruption of Mount Etna in 1754, a 19th-century wood-engraving.

tradition, dimly throw back at us the shadow of earliest human memory and explication?

Other gods and demi-gods are buried under Vesuvius and Etna, according to myth. Enceladus rebelled against the gods and is buried beneath Etna, while his brother, Mimas, was buried by Hephaistos beneath Vesuvius. In the *Aeneid*, Virgil's story of the origin of Rome written in the last two decades of the first century BC, we read one of the most dramatic accounts of an eruption in all classical literature. This is Virgil's account of Etna in full flow:

> The harbour there is spacious enough, and calm, for no winds reach it, but close by Etna thunders and its affrighting showers fall. Sometimes it ejects up to high heaven a cloud of utter black, bursting forth in a tornado of pitchy smoke with white-hot lava, and shoots tongues of flame to lick the stars. Sometimes the mountain tears out the rocks which are its entrails and hurls them upwards. Loud is the roar each time the pit in its depth boils over, and condenses this molten stone and hoists it high in the air.[6]

While volcanic activity provided a natural source for the imaginative myths of the ancient Greeks and Romans, it also drove early philosophers to attempt to explain what was going on so violently out of sight. Discussing the four great rivers of the earth, Plato writes in *Phaedo* of one of them, Pyriphlegethon, which

> pours into a huge region all ablaze with fire, and forms a lake larger than our own sea [the Mediterranean], boiling with water and mud; from there it proceeds in a circle, turbid and muddy, and coiling about within the earth it reaches the borders of the Acherusian Lake, amongst other places, but does not mingle with its water; then after repeated coiling underground, it discharges lower down in Tartarus; this is the river they name Pyriphlegethon, and it is from this that the lava-streams blast fragments up at various points upon the earth.[7]

Plato described the formation of lava or obsidian: 'Sometimes when the earth has melted because of the fire, and then cooled again, a black-coloured stone is formed.' Being cast into Pyriphlegethon is the fate that awaits 'those who have outraged their parents' – so watch out, kids. The Roman hero of Virgil's *Aeneid*, Aeneas, was shocked when visiting the Underworld to find Pyriphlegethon, 'sweeping round with a current of white-hot flames and boulders that spun and roared'.[8]

The Greek dramatist Aeschylus may have witnessed the 479 BC eruption of Etna, or at least some smoke and rumbling, when he visited the settlements of Magna Graecia in Sicily in the 470s. Indeed he died there in 456/455 when, according to legend, an eagle dropped a tortoise on his head from a great height. A contemporary of Aeschylus, the lyric poet Pindar, described Etna, 'from whose depths belch forth holiest springs of unapproachable fire'.[9]

The first recorded student of Etna was the fifth-century BC Greek philosopher Empedocles, who formulated the idea of the four elements, Earth, Air, Fire and Water. However, he used the mountain not as a laboratory to investigate the workings of the elements, but as a means to show himself to be the equal of the gods. There are variations in reports on the manner of his death: one tells that he wished to be immortal, and so threw himself into the crater. Unfortunately for his reputation one of his sandals slipped off as he fell, and this was found and gave the game away. Another legend tells that he believed he would return from the volcano as a god among men, while a third relates that he did indeed throw himself in, but was ejected during an eruption and landed on the moon, where he still survives by drinking dew.

Many classical philosophers grappled with the idea and purpose of volcanoes. Aristotle in *Meteorology* (fourth century BC) saw the earth as a living organism, subject to convulsions and spasms like any creature. He proposed that the fire beneath the earth is caused by 'the air being broken into particles which burst into flames from the effects of the shocks and friction of the wind when it plunges into narrow passages'.[10] He coined the word 'crater' ('cup' in Greek) to describe the dished form of volcano summits. Strabo in his *Geography* (first century AD) discussed the world's volcanoes, in particular those in and around the Mediterranean. He described Sicily as having been 'cast up from the deeps by the fire of Aetna and remained there; and the same is true both of the Lipari Islands and the Pithecussae' (Capri, Ischia and neighbouring islands).[11] 'Midway between Thera and Therasia', Strabo added, 'fires broke forth from the sea and continued for four whole days, so that the whole sea boiled and blazed, and the fires cast up an island which was gradually elevated as though by levers and consisted of burning masses.'[12]

Moving out beyond Europe, there are potent myths that weave in and out of the historical record. In the book of Psalms (18:7–8), God is described in terrifying terms:

> Then the earth shook and trembled; the foundations also
> of the hills moved and were shaken because he was wroth.
> There went up smoke out of his nostrils, and fires out of his
> mouth devoured: coals were kindled by it.

Salvator Rosa, *Death of Empedocles*, 1665–70, drawing.

This sounds like a graphic description of an earthquake and an erupting volcano. Sodom and Gomorrah are described as being destroyed by what looks like volcanic action:

> Then the Lord trained upon Sodom and upon Gomorrah brimstone and fire from the Lord out of heaven; And he overthrew those cities, and all the plain, and all the inhabitants of the cities, and that which grew upon the ground (Genesis 19:24–5).

The eruption of Hekla in 1104, in a 19th-century wood-engraving.

While Sodom and Gomorrah were real cities of the plain near the Dead Sea, and archaeological evidence has established that they were destroyed by a natural cataclysm around 1900 BC, this is likely to have been an earthquake rather than a volcano, as no volcanic activity has taken place in that region in the past 4,000 years. Muddled superstition and dogma led to the obvious conclusion that volcanoes marked the entrance to hell. Following Plato, and in commentary on the Book of Revelation, St Augustine wrote in *City of God* of hell as having 'a lake of fire and brimstone',[13] while variously Etna and Vulcano were both believed to be mouths of hell. Drawing boundaries out into the cold north, the Cistercian priest Herbert of Clairvaux opined, after its 1104 eruption, that Hekla in Iceland was the mouth of hell. This story was repeated again and again until Jules Verne used Hekla and another Icelandic volcano, Snaefells, as the gateways to the centre of the earth in *Journey to the Centre of the Earth* (1864).

The embryonic Icelandic parliament, the Althing, met in AD 1000 at Thingvellir, a volcanic cliff with remarkable acoustic properties some 48 kilometres (30 miles) northeast of Reykjavík.

The wide plain that stretches away from Thingvellir marks the line of the geological fault dividing the North American and the Eurasian plates. Running approximately down the centre line of the Atlantic Ocean, it passes through or near the Azores, Ascension island and Tristan da Cunha. Disputing parties met at Thingvellir to reach a decision about the religion of Iceland: to choose between worship of the old Nordic gods, or the new Christianity. During the interminable meeting a messenger brought news that fire and molten rock were erupting from the ground near the village of the chief Nordic advocate, and threatening destruction. Followers of the local pantheon interpreted this as meaning that the gods were angry at the proposal. But wait, said the advocate of Christianity, 'with whom are the gods angry? They cannot be angry at what is not there, as there is no Christianity in the island. They are angry, instead, with the old ways.' That turned the vote, and the Christian lobby won the day.

William G. Collingwood, *Thingvellir, Meeting of the Althing, Iceland*, *c.* 1900, gouache on card.

Across the globe, the volcanoes in the Pacific are the home of the goddess Pelé. Born in Tahiti, Pelé was chased after falling out with her sister Namakaokahai, 4,000 km (2,500 miles) away to the Hawaiian archipelago. As she swam from island to island, northwest to southeast, Pelé left craters and mountains behind her – Diamond Head, on the island of Oahu; Haleakala, on Maui; and Kilauea, on the island of Hawaii. This progression matches the modern discovery that the volcanoes become younger towards the southeast. Pelé's flight ended in Hawaii, where she created Halemaumau, the crater of Kilauea. There, according to the legend, she still lives and causes eruptions. Pelé has a short temper, and can open up craters with a kick of her heel, hurling lava about wildly. But before each eruption, she is said to give some warning by appearing either as an old woman, or a beautiful girl. Pouring as it does from a crack in the earth, Hawaiian lava has created a vaginal form for its crater. Thus local tradition has it that an eruption is a sign of the menstruation of Pelé, in which the lava flows to the sea, the place of ritual cleaning.

Kilauea in eruption, late 19th-century lithograph.

Lava flow in Hawaii pouring into the sea.

Every year Hawaiians gather in ceremonial costume at the edge of Kilauea to honour Pelé. In 1824 Kapiolani, the Christian wife of an island chief, tried to enrage the goddess by throwing stones into the crater. Pelé remained calm, and this persuaded many to convert to Christianity. When a lava flow threatened the city of Hilo in 1881, Hawaii's indigenous religion had been replaced almost totally by Christianity. Awareness and respect for Pelé was still widespread, however, and an appeal was made to the Hawaiian princess Ruth Keelikolani to intercede with Pelé. Ruth did so, and the flow stopped just outside Hilo.

The Native Americans of Oregon believe that a destructive fire god lived in Mount Mazama, and a beneficent snow god in nearby Mount Shasta. The two struggled, and the snow god won, decapitating its enemy. To mark this victory of good over evil, the crater of Mount Mazama filled with water. Formed

6,000 years ago, this may make it the second earliest eruption recorded in legend, after Çatal Hüyük. Shasta was considered to be the centre of creation, the Great Spirit having created it out of ice and snow from heaven. Using the resulting heap as a stepping stone, he created the flora and fauna of the earth. Shasta retains its religious significance, being the base for dozens of New Age sects, an alleged UFO landing place, and the way into the fifth dimension. It is also a very popular ski resort.

Further north in the Cascades chain is Mount St Helens, which erupted with such violence in 1980. This too enters Native American legend, which explains something of its eruptive nature, and that of its neighbouring volcanoes. Legend follows the typical pattern of quarrels among gods or spirits over land or love, in the case of the Cascades range the love being that for the beautiful Loowit, who was fought over by two young Indian nobles, Pahto and Wy'east. Becoming furious at the devastation the quarrel had caused – the shaking of the earth and the creation of the Cascades – the chief of the gods struck his sons down. Pahto became what is now Mount Adams, Wy'east became Mount Hood, and Loowit was transformed into Mount St Helens, originally 'Louwala-Clough', translated as 'Smoking Mountain'. And so the legends go on. In Wyoming, the Devil's Tower, a column of lava isolated by erosion, has vertical striations believed to have been the claw marks of a bear chasing young girls.

Other legends from the Americas include the Peruvian story of the volcano Misti, near Arequipa, which was plugged with ice by the sun god as a punishment for pouring lava across the landscape. Bringing legend forward into the modern world, Huaynaputina, a neighbouring volcano to Misti, erupted in AD 1600 and covered Arequipa with ash. At this turbulent interface between religions, Christians and pagans alike feared the end of the world, and human and animal sacrifices were made in appeasement. But why had Misti not erupted in sympathy, the Christians asked? The reason they gave was because the Spaniards had baptised Misti and renamed it San Francisco. Paganism and Christianity made equal use of volcanic activity to defend their

Mount St Helens erupting in 1980.

Ol Doinyo Lengai
volcano, Tanzania.

beliefs: Aztecs claimed in the sixteenth century that eruptions of Popocatepetl were the response of local gods to the Spanish conquistadors profaning their temples. Over 300 years later, the natives of Lake Ilopango in El Salvador attributed the 1879–80 eruption of the Islas Quemadas volcano to the anger of the goddess of the lake at the introduction of a government steamboat service. To appease her, the local population fled back to time-honoured traditional practice: they bound a young child hand and foot and threw him into the lake.

The Swahili Kilima Njaro is translated as 'Shining Mountain', an appropriate name for Kilimanjaro, at 5,895 metres the world's highest single mountain, unconnected to a range. This has not erupted in recorded history, although its neighbour in the African Rift Valley, Ol Doinyo Lengai, is active, having erupted many times in the last century. Its name in the Masai language means 'Mountain of God', and far from being a source of terror, it is seen as a bringer of fertility and bounty. When it erupts, nursing mothers express their breast milk on its flanks in gratitude.

In the North Island of New Zealand, according to Maori legend, Taranaki (also known as Mount Egmont) and Ruapehu both fell in love with Tongariro, present-day Ngauruhoe. Taranaki attacked the volcano Ruapehu who, in response, poured showers of boiling water from his crater lake. Taranaki then erupted stones that destroyed Ruapehu's cone. Ruapehu swallowed and spat them out over Taranaki, who fled out to sea down the valley of Wanganui. The Maori still do not bury their dead on the line between Egmont and Ruapehu, in case they start fighting again. When another North Island volcano, Tarawera, erupted in 1886 three villages were buried and more than 300 people died. The Maori believed that the villagers had either eaten wild honey, an act that was taboo, or had become degenerate through contact with Europeans.

These and other myths from volcanic parts of the world revolve around love and hate, peace and war, mercy and punishment, religion and superstition. All are fundamental qualities in human nature that have independently developed to counter, explain and try to quell what is a simple and unavoidable physical fact: that the heat and pressure within the earth's core will in its own time burst through weak seams in the crust and pour its contents across the landscape. It was not until around AD 580 that the connection between volcanic output and the fertility of the land was described in surviving literature. In the warmth and relative comfort of Caesarea, modern Israel, the historian Procopius noted that 'after Vesuvius had spat out its ashes, the harvests of the neighbouring countryside were abundant'. This is perhaps the earliest known observation in European literature of a volcano that has a positive, even encouraging tone.

2 The Geography, Science and Allure of Volcanoes

Looking at an old painted canvas, one can sometimes find distinctly outlined areas of cracking, or 'craquelure', as it is known in the art trade. The cracks make irregular and varied patterns across the paint surface, creating distinct divides from one area to the next. As the paint dries, so the gaps between these areas widen, until all movement ceases.

So, to a very minor extent, with the surface of the Earth, the lithosphere. In its irregular cooling over thousands of millions of years the lithosphere, land and sea, has cracked into what has become an established pattern – seven large plates that enwrap the globe: the North American, Eurasian, Pacific, South American, African, Indo-Australian and Antarctic plates. These have travelled the globe for many millions of years, since the crust first developed a rigid surface roughly 2.5 billion years ago. In addition, seven smaller plates range in area from the largest, the Nazca plate in the southeast Pacific, to smaller ones such as the Cocos and Juan de Fuca plates. These are at the Pacific coasts of Central America and parallel to the Cascades range in North America respectively. The lithosphere lies like a skin of paint on top of the asthenosphere, an inner layer of the earth that is comparatively weaker than the lithosphere and can flow over time, allowing the lithosphere some degree of surface movement. It is this freedom that causes the lithosphere to adjust itself continually in relation to the earth's inner energy, as a coat jacket will adjust itself in response to the movement of the shoulders of its wearer. Plato described

San Andreas Fault, California, 1992

the earth beautifully in *Phaedo*; he might have been writing of tectonic plates:

> The earth, viewed from above, looks like one of those balls made of twelve pieces of leather, painted in various colours ... Up there the whole earth displays such colours, and indeed far brighter and purer ones than these. One part is marvellously beautiful purple, another golden: the white is whiter than chalk or snow ... more beautiful than any we have beheld.[1]

Thus the movement of the lithosphere continues, as the earth cools and as plates change their dimensions relative to each other. The size and number of plates has adjusted over the millennia, and continues to change: the Indo-Australian plate is understood to be very slowly coming apart to form the distinct Indian and Australian plates. Other much smaller plates, like flakes of damaged paint on a canvas, lie unstable beside larger ones, in particular in the area around Indonesia and the Philippines. And how slow the changes are: the rates of movement of the plates across the globe vary from the speed of growth of the human fingernail, to that of human hair. Very slow indeed; though one might reasonably feel that it is not slow enough.

In most cases volcanoes emerge along the lines of the meeting of the plates. Thus commuters travelling to work in Birmingham, England, or Alice Springs, Australia, will not encounter a volcano, as they are thousands of miles from a tectonic plate boundary. Commuters in Naples, Reykjavík or Mexico City, however, are at a statistical disadvantage here, while nearby plate boundaries move together or draw apart millimetre by millimetre.

The earth's volcanoes, principally, run along the lines known as the Pacific Ring of Fire, which edges the Pacific and Nazca plates, and the mid-Atlantic oceanic ridge. This is a crack in the earth's surface running north–south from beyond Iceland to Antarctica, with vents all along it like a line of buttons on that jacket: as the wearer of the jacket puts on weight, so the buttons begin to pull and fly off. In the case of the earth, it is not

Lava flow from Arenal Volcano eruption, Costa Rica.

the putting on of weight that is the cause of undersea volcanic eruptions, but relentless pressure of heat from the planet's interior. Further spurs of volcanoes run in a line south along the East African Rift Valley from the Red Sea – itself the product of the divergence of the African and the Arabian plates – and from Turkey and Greece to southern Italy, where the African plate pushes up against the Eurasian plate. Yet other volcanoes crop up where the lithosphere happens to be thin, at great distances from plate boundaries, as in Hawaii, Central Asia and West Africa.

As the tectonic plates move in relation to each other, so they perform different approaches. They diverge, or pull apart, as in the mid-Atlantic ridge, of which Iceland is a volcanic effusion. They rub together laterally, as at the San Andreas Fault in California. A third meeting is convergence and collision. In this case, there is a further alternative: either one plate moves below the other, as in the eastern Pacific, creating a subduction zone along the active length, the Andes range of mountains; or they collide like clashing cymbals in what is plainly known as a 'continental collision'. This has created other mountain ranges, such as the Himalayas or the Alps, by a crumpling effect.

Volcano Laki, Iceland, fissure vent with spatter cones.

The majority of the volcanoes on land, which are the subject of artists' observation and inspiration, are the result of plates moving towards each other and causing a subduction zone. Examples of these are in the west coast of South America, where the Nazca plate runs up against the South American plate, and in Europe where the African plate presses the Eurasian. These movements over millions of years created the volcanic landscape of South America, and the volcanoes that pepper the line from eastern Turkey to central Italy.

It remains an irony that an atypical volcano such as Vesuvius should have attracted so much attention, but this is only because the lonely spur of which it is a part happens to be at or near the cradle of Western civilization. Dramatic though Vesuvius certainly is when in eruption, it is the thousands of other volcanoes that erupted unseen and unrecorded by the literate West until well into the twentieth century that demonstrated the sheer energy and destructive power of the earth's magma welling up through the cracks in its surface. Over the wastes of time these have piped out into the heavens, unseen and unheard, like a slow, mournful concerto for organ. To imagine Italy's Vesuvius as representative of the totality of the earth's volcanoes would be comparable to regarding a comedy such as *Two Gentlemen of Verona* as representative of the range and mastery of Shakespeare's plays; or a Ferrari, beautiful though it may be, as the only car worth talking about.

Volcanoes fall into a number of types, many of which are depicted by the artists whose work will be discussed in these pages. A volcanic 'fissure vent', as depicted by Finnur Jónsson, is a typical feature of the Icelandic landscape, where the faultline in the earth pulls slowly apart, releasing the energies within. In the Icelandic landscape such activity produces low-lying broken hill forms, characterized by blackened lava-rich earth, surface

Finnur Jónsson, *Lakagígar Craters*, 1940, oil on canvas.

Brynjólfur Thórdarson, *Hekla*, 1939, oil on canvas.

Charles H. Owens, *Kilauea Crater, Hawaii*, 1912, gouache, pastel and watercolour.

cracks and upheaved rock formations. Iceland is also the home of 'shield volcanoes', as is Hawaii, where low-viscosity (comparatively runny) lava wells up from a vent, cooling layer upon layer to form a low-lying dome with a profile like an ancient shield. Works by Brynjólfur Thórdarson of Hekla, and Charles Owens of Kilauea, are representative of these.

A third type is a stratovolcano, with its elegant conic form, which has come to represent the beauty and allure of the earth's volcanoes. These have been built up by layer upon layer of lava alternating in strata – hence its name – with ash or other ejected material. One of the finest examples of a stratovolcano is Mount Fuji in Japan, as depicted in thousands of Japanese prints, notably by Hokusai and Hiroshige in the nineteenth century and by David Clarkson in our own day. However, to make the volcano seem yet more elegant in form, Japanese artists tended to exaggerate the angle of its slope to above 45°, though the actual angle of the sides of Fuji is no more than 30°. Vesuvius is another stratovolcano, but its pure shape was damaged around 18,000 years ago when its top was blown off in a very large eruption that threw it down to form 'Monte Somma', the rocky collar of Vesuvius to its northeast. The present active summit and cone of Vesuvius has grown up over the intervening millennia.

overleaf:
Utagawa Hiroshige, *Musashi Koganei*, from *Thirty-six Views of Mount Fuji*, 1858–9, woodblock print.

冨士三十六景

David Clarkson, *Fuji AM* (top) and *Fuji PM*, 2002, acrylic on canvas.

Lava domes, like shield volcanoes, are formed by the slow eruption of very sticky, or viscous, lava that extrudes within a volcanic landscape. Within the crater of an existing volcano they can appear like a bubble. After the eruption of Mount St Helens in 1980, lava domes emerged within its crater, revealing ongoing volcanic activity deep within the mountain. Like a lava dome, a 'hornito' is a bubble of upswelling lava that has emerged from a larger flow to create a modest-sized hillock of ash or volcanic rock. Named after the small Spanish oven that their shape resembles, hornitos are common in, for example, Lanzarote in the Canary islands. It was there that the Scottish artist Wilhelmina Barns-Graham (1912–2004) discovered hornitos as subjects in the early 1990s, and gave expression to their irregular regularities and gloominess within an exuberant sub-tropical landscape.[2] Greatly absorbed by the slow-flowing contours of hardened jet-black lava, Barns-Graham made drawings and paintings in which her naturally fluid line moves in mild curling parallel

Wilhelmina Barns-Graham, *La Geria, Lanzarote No. 3*, 1989, acrylic on paper.

Ilana Halperin, *Physical Geology, No. 2*, 2008, C-print.

across the page. She depicts the hornitos as central, black, rigid forms richly contrasted by the high colour of flowers, fields and houses. We are in the midst of lush dampness in Barns-Graham's paintings, with aridity and harsh reality just a cricket's hop away.

The appearance of volcanoes, and thus their appeal to artists, depends entirely on the type of magma from which they were originally formed. As a general rule, the runnier the lava, the flatter the form of the volcano in the landscape – this is obvious. Low-viscosity lava is low in silica and high in basalt content: the Hawaiian volcanoes emit low-viscosity lava, resulting in pools and lakes, and long streams which, in Big Island, flow into the sea, creating new land. This activity has been the subject of the work of Ilana Halperin. The volcano Nyiragongo, on the borders of the Democratic Republic of Congo and Rwanda, has similarly fluid lava that collects in a crater lake at the summit of the volcano. This highly unstable mass erupted most recently in 1977 and 2002, when the walls of the crater collapsed, allowing the lava

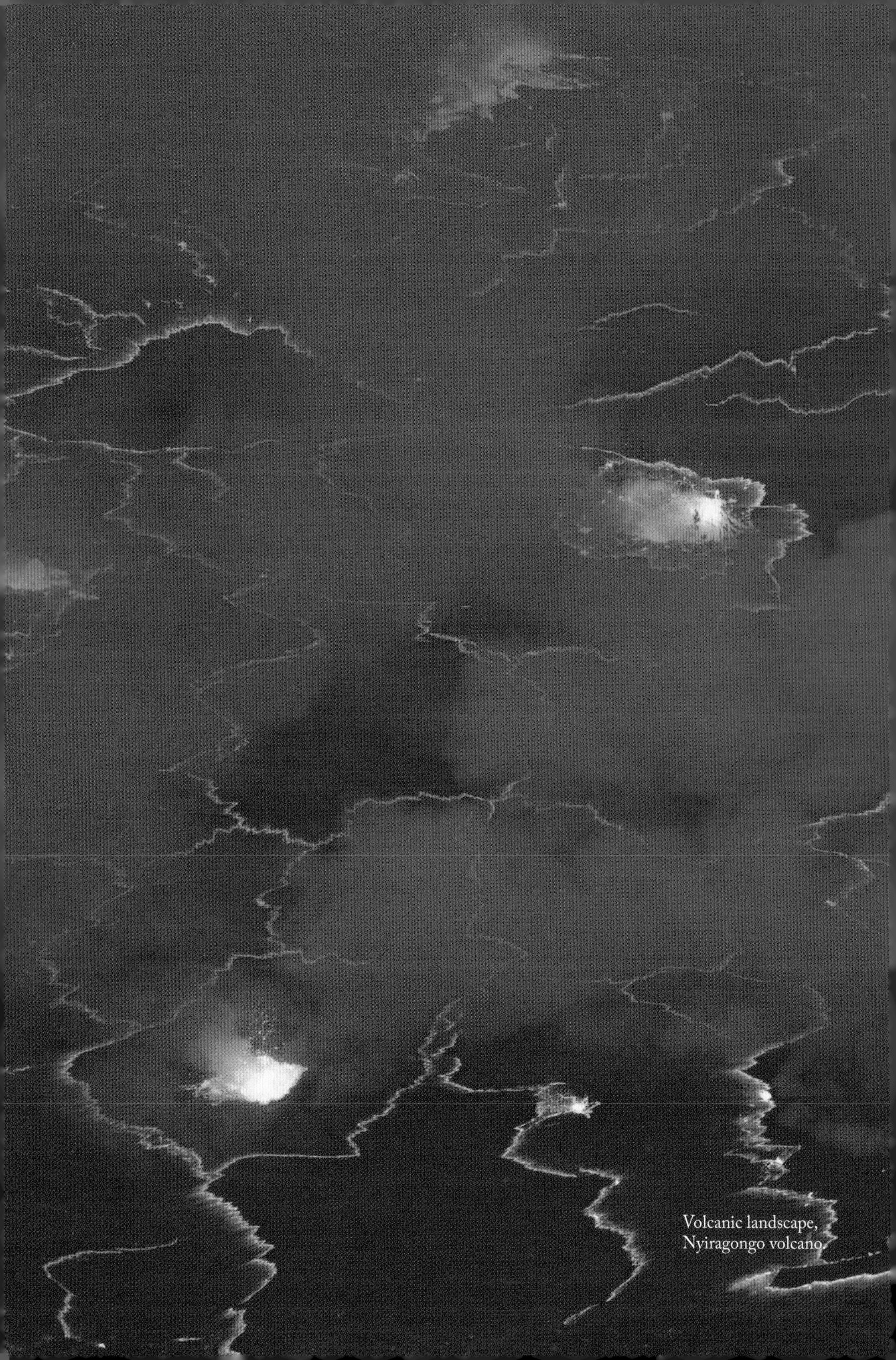

Volcanic landscape,
Nyiragongo volcano.

A. Halut, *Eruption of Nyiragongo*, 1977, oil on canvas.

Lava in Hawaii pouring into the oceans.

to flood into the valley beneath. There were many casualties and much damage to buildings, crops and livelihoods. The artist Halut was present during the 1977 eruption, and was among the many refugees.

Vesuvius, whose lava has a relatively high silica content, produces slow-flowing pyroclastics, the hot mixture of materials erupted from the interior. This gradualism has the tendency to cause the eruptions to be more powerful, and to eject to greater altitudes, as happened during the AD 79 eruption, witnessed and described by Pliny the Younger. In that event a column of fiery lava drove skywards for three or more kilometres. The thicker the lava, as at Vesuvius, the longer the eruption column holds its shape; the thinner it is, on the other hand, the more it spatters and fountains, as is characteristic of the Hawaiian volcanoes. The author Norman Lewis was present at the 1944 eruption of Vesuvius: 'At the time of my arrival the lava was pushing its way very quietly down the main street . . . I had been prepared for rivers of fire, but there was no fire and no burning anywhere –

View of the great eruption of Vesuvius in 1872, wood-engraving.

only the slow deliberate suffocation of the town under millions of tons of clinker . . . the whole process was strangely quiet.'[3] By contrast, the constant flow of lava in Hawaii was described by Ilana Halperin in 2009 thus: 'The amount of lava that has come out of the Kilauea volcano since 1984 . . . could pave a road to the moon and back again at least five times . . . Lava moving as if a normal body of water – like the Hudson, only blood red.'[4]

The terms given to volcanoes to describe their current state are active, dormant (or quiescent) and extinct. None of these terms are wholly satisfactory, as volcanoes exist in a time frame so much longer than that of humans. While 'active' is clear enough, 'extinct' may be a matter of opinion, or of hope. Volcanoes thought to have been extinct for thousands of years have suddenly come to life, indicating that we just do not know enough about them. An example here is Fourpeaked Mountain in Alaska, thought to have been extinct for 10,000 years, which erupted in 2006. Thus it was dormant all along, not extinct. The same for Vesuvius: Pliny the Elder gave no consideration to Vesuvius in his account of the world's volcanoes in *The Natural History*, but then it killed him. We can be certain about volcanic extinctions such as those in Edinburgh, where Arthur's Seat is a 350-million-year-old volcanic plug, well worn by later glaciations, and no longer plugging back any hot eruptive material. We can also be sure that the volcanic activity that formed the Northwestern Hawaiian islands, the tail of the chain of islands pointing northwest, ceased more than seven million years ago. This is because the thinner part of the lithosphere that created the 'hotspot', which allowed them to be formed has moved away from them with the motion of the Pacific plate, to enable the creation of Hawaii's Big Island. By contrast, the hotspot from which Iceland emerged on the mid-Atlantic ridge about 20 million years ago – long after the extinction of the dinosaurs – is static in relation to the divergent tectonic plates. For this reason the island remains actively volcanic.

A type of volcano that can only be imagined by artists is the supervolcano. These lie, neither active, dormant nor extinct, beneath the lithosphere in places including Yellowstone in the USA, the Campi Phlegraei near Naples, Sumatra and New Zealand. A relatively recent taste of their power was experienced in 1815 when Tambora erupted in the Philippines, and in 1883 when Krakatoa erupted. Volcanologists today may be able with sophisticated equipment to forecast volcanic activity and try to evacuate local populations, but there is nothing that they or any human beings can do to stop eruptions. We are living on the top of the furnace that supports us, another of the exciting and unpredictable aspects of being human.

3 'A Horrid Inundation of Fire'

The first coherent list of active volcanoes was made by the Roman natural philosopher and historian Pliny the Elder (AD 23–79) in his *Natural History*. Pliny wrote of Etna, which 'always glows at night'; of Chimaera in Phaselis (now in Turkey), which 'blazes day and night with a continuous flame'; of Hephaestus in Lycia, which 'flares up so violently when touched by a blazing torch that even the stones and sand in rivers glow'. Of Stromboli he wrote that 'it differs from Lipari only in the flame of its volcano being brighter'.[1] One volcano he did not mention because it was not considered a danger – Vesuvius – was to bring about his death.

The terrible event of 24 August AD 79 was described in the greatest possible detail by Pliny's nephew, known to us as the letter-writer Pliny the Younger (62–114). His is usually said to be the first eyewitness written account of an eruption, though the description of Etna exploding, written by Thucydides who was in Sicily at the time of the 400 BC event, clearly has the ring of the eyewitness. The seventeen-year-old Pliny, studious and informed, later described what he could recall of the Vesuvius eruption in two letters to the historian Tacitus. It was an extraordinary good fortune for posterity that he should be present at one of the formative events of European culture: the right young man, in the right place at the right time, looked on from Misenum across the bay as Vesuvius erupted, and described what began as an ordinary day and ended in mass death. His uncle, who was commander of the Roman fleet based at Misenum, was with

George Poulett Scrope, *Eruption of Vesuvius as seen from Naples, October 1822*, 1823, lithograph.

him, and seeing the huge column of smoke insisted on being rowed across the bay of Naples to Pompeii to take a closer look. His voyage ended in his choking to death from gas and ash inhalation, while his nephew had the safer, longer view. But before he set off, uncle and nephew saw the beginning of the action together. It was early in the afternoon. Significantly, they could not at first make out which mountain was erupting: 'it was not clear at that distance . . . it was afterwards known to be Vesuvius'.

> Its general appearance can best be expressed as being like an umbrella pine, for it rose to a great height on a sort of trunk and then split off into branches, I imagine because it was thrust upwards by the first blast and then left unsupported as the pressure subsided, or else it was borne down by its own weight so that it spread out and gradually dispersed. Sometimes it looked white, sometimes blotched and dirty, according to the amount of soil and ashes it carried with it.[2]

There seems to have been no sign of fire at this stage. Nevertheless, the elder Pliny gave orders for his fleet to be launched on a courageous mission to rescue the people on the other side. His nephew watched as they rowed away:

> Ashes were already falling, hotter and thicker as the ships drew near [Pompeii], followed by bits of pumice and blackened stones, charred and cracked by the flames.

As darkness fell, the flame really became apparent: 'Meanwhile . . . broad sheets of fire and leaping flame blazed at several points, their bright glare emphasised by the darkness of night.' Then the earthquake took hold: 'Buildings were now shaking with violent shocks, and seemed to be swaying to and fro as if they were torn from their foundations.'

By the time he came to write the letters, Pliny the Younger had been able to gather up the bits of scattered information about his uncle's death, and the deaths of thousands of others:

> Then the flames and smell of sulphur which gave warning of the approaching fire drove the others to take flight and roused [my uncle] to stand up. He stood leaning on two slaves and then suddenly collapsed.

Finally, the death:

> The dense fumes choked his windpipe which was constitutionally weak and narrow and often inflamed. When daylight returned on the 26th – two days after the last day he had seen – his body was found intact and uninjured, still fully clothed and looking more like sleep than death.

In his second letter to Tacitus, Pliny described a further phenomenon, the sea drawing back: the first sign, as we now know, of a great wave or tsunami:

> We also saw the sea sucked away and apparently forced back by the earthquake: at any rate it receded from the shore so that quantities of sea creatures were left stranded on dry sand.

Ending his second letter to Tacitus, Pliny the Younger described the aftermath of the eruption:

> At last the darkness thinned and dispersed into smoke or cloud; then there was genuine daylight, and the sun actually shone out, but yellowish as it is during an eclipse. We were terrified to see everything changed, buried deep in ashes like snowdrifts.[3]

This is as precise and as particular a description of the event as any we could wish for now. It is pre-scientific, as well as scientific, and relies on observation without the complication of too much knowledge. Pliny's letters, discovered in the sixteenth century, give an extraordinarily clear account of the chronology of a Vesuvian eruption, and came to form a convenient basis for later accounts and writing, in particular Edward Bulwer Lytton's

Fresco from Pompeii, before AD 79. Bacchus is standing before Vesuvius, which had yet to erupt.

Last Days of Pompeii, written nearly 1,800 years later. They do not however tell us anything about the appearance of Vesuvius before the eruption. Evidence for this was buried by the ash that settled on the city, and was not dug out again until the nineteenth century. In this section of fresco, found in Pompeii, Bacchus, the god of wine and good living, holds in his right hand a *kantharos*, a high-handled cup, that he is casually allowing to empty. His accompanying leopard is ready to lick up the drops of wine. The body of Bacchus is formed out of a bunch of grapes, which suggests that he is many-breasted, or at least fruitful, like the Syrian god Cybele. Behind him is the source of the grapes, a mountain whose lower slopes are covered in vines. This has generally been interpreted as Vesuvius, the dominating feature of that landscape, which was particularly fertile on account of its volcanic soil.

The historian Strabo described Vesuvius before the eruption changed its shape for ever:

> Above these places [Pompeii, Herculaneum] rises Vesuvius, well cultivated and inhabited all round, except its top, which is for the most part level and entirely barren, ashy to the view, displaying cavernous hollows in rocks, which look as if they had been eaten by fire, so that we may suppose this spot to have been a volcano formerly, with burning craters, now extinguished for lack of fuel.[4]

Vesuvius is one of the most frequently altered mountains in general knowledge: its largest eruptions have had significant effects on its height, form and shape at its summit. If it is Vesuvius that is represented in the fresco, and it is reasonable to suppose that it must be, it is shown in its rugged conical form before the AD 79 explosion, but long after the immense eruption around 16,000 BC ripped off its top and formed Monte Somma. The AD 79 eruption lowered what was left of the mountain and flattened its top, allowing the summit to flux and change, eruption by eruption, until the most recent event in 1944.

Meanwhile, 270 km (170 miles) to the south, Etna shivered and shook as Typhon and Enceladus variously scratched themselves and writhed unhappily within. There had been a large eruption of Etna in AD 38–40 that was heard all round Sicily, but the mountain has only once, in 1669, produced an event quite as spectacularly violent as Vesuvius in 79. Etna is of a different formation to Vesuvius, having vents all around its flanks that are almost constantly in low-level eruption, making it the most active volcano in Europe.[5]

For those relatively few people in Europe in the Middle Ages who knew of their existence, active volcanoes were sources of natural terror. A volcano is usually visible to residents of its

Fig.1. Vesuvius in the Pre-Historic Period
Fig.2. Vesuvius in the Classical Period

Vesuvius in the prehistoric period and Vesuvius in the classical period, shown in a late 19th-century lithograph.

region; coming earthquakes, of course, are not. Those few literate witnesses of volcanoes in action have left patchy but colourful evidence. The twelfth-century monk Benedeit, who told the story of St Brendan's voyage six hundred years earlier, described Hekla erupting: this may have been the 1104 eruption. The monk sees Iceland as

> a smoky and foggy land, stinking worse than carrion . . . throws fire and flames, blazing beams and scrap iron, pitch and sulphur up to the clouds, then everything falls back into the abyss . . . Judas was imprisoned there.[6]

J.-B. Masculus, The 1631 eruption of Vesuvius with St Januarius, an engraving of 1633.

According to the Icelandic saga *The Flatey Book*, eyewitnesses of the 1341 eruption of Hekla saw birds that they took to be human souls flying into the erupting fire.[7] Iceland was cold and distant, the home of a mysterious people who spoke an incomprehensible language. From a European perspective, Iceland was not on the way to anywhere. If travellers sought warmth, comfort and trading opportunities in the Middle Ages, they would more profitably travel east and south in Europe, rather than north and west. The Icelandic Chronicles, according to Uno von Troil, later the Archbishop of Uppsala, who climbed Hekla with the young English natural philosopher Joseph Banks in 1772, listed 63 eruptions in Iceland between 1000 and 1766, 23 of them of Hekla.[8]

When medieval historians and natural philosophers wrote about volcanoes, it was either to describe them or to try to divine some kind of understanding of how they worked. The Oxford scholar Alexander Neckam (1157–1217) first appears to have used the word 'Volcano' to describe a place where the earth's fire burns, though in this instance he used the word as the dative case of *vulcanus*, meaning '[gives] to Vulcan':

> If however you think that earth is black, while the three other elements [air, fire, water] are bright, there will be others who regard them as the same. Why? They reply: vision gives to Vulcan the glow that is inherent, but what does that glow mean? The powerful force of Nature, the friendly companion of things, binds together matter and fire.[9]

What he seems to be trying to say is that all the elements are bound up together in nature, which is proved by the glowing of Vulcan's mountain.

The German scholar saint Albertus Magnus (*c.* 1200–1280) made the first experimental model of a volcano, using an enclosed brass vase with two stoppers. Filling this with water and bringing it to the boil, he found that the pressure inside made either the upper cork fly out, followed by a plume of steam, or the lower cork, followed by a fountain of boiling water. This experiment, one of the earliest recorded, was fully in line with Albertus's assertion that 'natural science does not consist in ratifying what others have said, but in seeking the causes of phenomena'. These words are direct precursors of the motto of the Royal Society, *nullius in verba*, which means 'do not take anybody's word for it', or 'find out for yourself'. That is what the first secretary of the Royal Society, Henry Oldenburg, tried to do when he wrote in 1668 to Thomas Harpur, a resident of Aleppo, Syria, on behalf of the Society. He sought answers to some perplexing questions about the geology and geography of Asia Minor, and asked Harpur 'whether Mount Sinai is known ever to be a volcano', and 'whether there be any volcano in your parts, or nearby'.[10]

The event that may have prompted Oldenburg's question was the coming English-language publication in 1669 of *The Vulcanos: or, Burning and Fire-Vomiting Mountains*. This was a translation of parts of *Mundus Subterraneus* by the German Jesuit Athanasius Kircher (*c.* 1601–1680), first published in Latin in 1665. Kircher was driven by the admixture of extraordinary genius and religious obligation to become the most learned and active savant of his age. While he may not, as traditionally claimed, have been the last man to know everything, he did

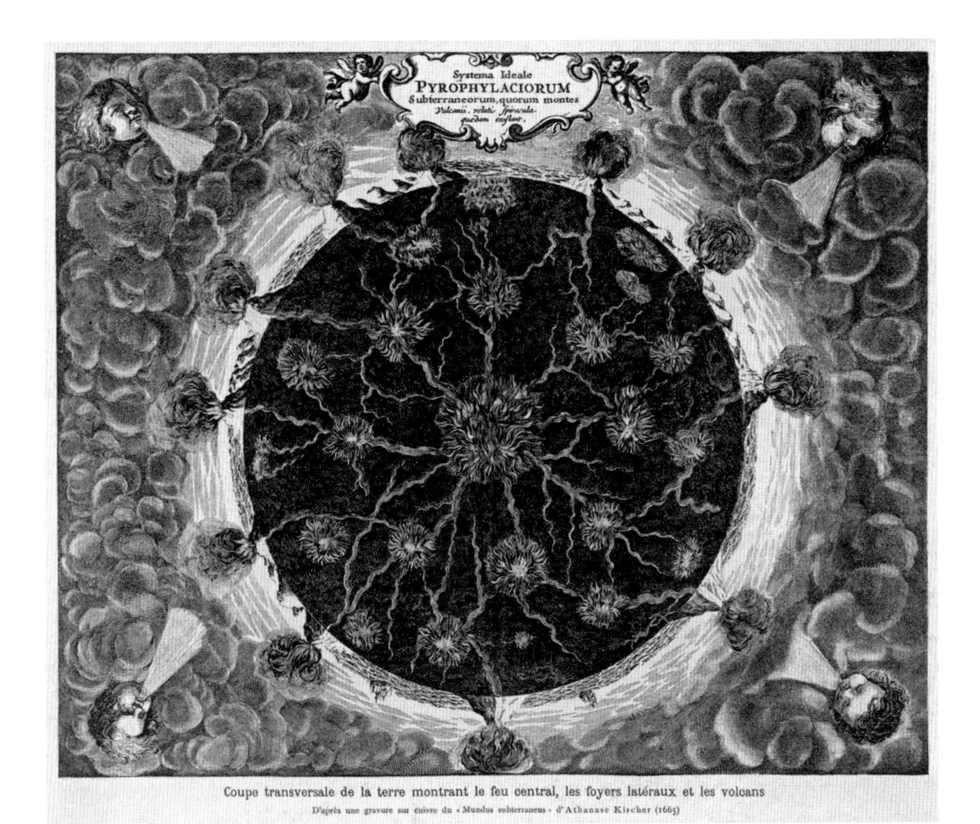

Systema Ideale
PYROPHYLACIORUM
Subterraneorum, quorum montes
Coupe transversale de la terre montrant le feu central, les foyers latéraux et les volcans
D'après une gravure sur cuivre du « Mundus subterraneus » d'Athanase Kircher (1665)

An engraving in Athanasius Kircher, *Mundus Subterraneus* (1665), showing a cross-section of volcanic channels within the earth, as understood in Kircher's day.

hold the world's knowledge in his hands and cherished it all, publishing on every subject under the sun.

In the chaos of northern Europe shattered by religious and military strife, Kircher led a charmed life that spanned the Thirty Years War and the Counter-Reformation. He had not only the intellectual capacity but also the organizing genius to prospect a route through knowledge and its accumulation, to its expression and distribution. Kircher wrote books on Egyptology, geology, medicine, mathematics, and on China. His mind and intellectual ambition embraced so wide a landscape that he was in demand at the courts and universities of France, Germany and Austria, as well as at the Church in Rome.[11]

While volcanology was only one of the topics covered in *Mundus Subterraneus*, along with the working of the tides, the

Athanasius Kircher, *Mount Etna erupting in 1637, as seen by the author*, 1637.

weather, fossils and early man, it is Kircher's understanding of volcanoes and the illustrations of them that particularly caught the imagination of the fellow scholars and the narrow band of literate Europeans in his day. The English edition extracted the chapters on volcanoes, and added 'a scatter'd Collection of Historical Relations by others, of most remarkable passages'. Published as it was three years after much of London was burnt to the ground in 1666, it was prefaced by a poem that attempted to put the capital's recent agonies into perspective:

> None sadlier knows the unresisted Ire,
> Then [*sic*] Thou, Poor London! of th'all raging Fire.
> But these occasion'd kindlings are but Blazes,

> To th'mighty Burnings, which fierce Nature raises.
> If then a Town, or Hills blaze be so dire;
> What will be th'last, and Universal Fire?

The Vulcanos: or, Burning and Fire-Vomiting Mountains, the first modern account in English of volcanoes across the globe, sets out the state of knowledge in a pocket-sized volume.[12] Oldenburg's question to Harpur may have been one attempt to ensure that the facts in the book were correct. Its publication was clearly serendipitously timed, appearing as it did soon after the catastrophic eruption of Etna in March 1669. Kircher's worldview was maintained in the English version, which was liberally extended from the original by other accounts and amendments:

> The Atlantick Sea so abounds with subterraneous Fires, that Plato's Land, call'd Atlantis, seems to have been swallow'd up from no other cause, but the outrages of these fires and earthquakes thence arising . . . Yet no part of the world is more famous than America, which you may call Vulcan's Kingdom. In the Andes alone, which they call the Cordillera from a Concantenation of mountains in the Kingdom of Chile are fifteen Vulcanos, Terra del Fuego. In Peru not fewer than in Chile; six of inaccessible height; and three in the continued tops of the Andes, besides innumerable Vulcanian Ditches, Pits, and Lakes . . . In the Northern America, are observed five, partly in new Spain, viz. three, formidable for their belching flames, partly in new Granada, partly in the very heart and midst of California, and the more inland Mexican kingdom.[13]

The book goes on to describe many other volcanoes all over the world, including those in Persia, the Asian steppes, Mollucca (the Spice Islands) and the Philippines, Sumatra, Japan, Tenerife and St Helena. Following Albertus Magnus, Kircher's central task for his readers was to try to demonstrate with engravings and text how volcanoes work:

> We would . . . hereby shew, that the bowels of the Earth are full of Æsturies, that is places overflown, and raging with Fire, which we call underground Fire-houses, or Conservatories; whether after such, or in any manner disposed. From the Centre, therefore, we have deduc'd the Fire, through all the Paths (to be supposed) of the Terrestrial World; even to the very Vulcanian Mountains themselves, in the Exteriour Surface.[14]

As a courageous example of extreme information-gathering, Kircher had himself lowered into the heaving red crater of Vesuvius at night in 1638, during one of its actively threatening periods. His report is graphic in the extreme:

> I saw what is horrible to be expressed, I saw it all over of a light fire, with an horrible combustion, and stench of Sulphur and burning Bitumen . . . Methoughts I beheld the habitation of Hell . . . An unexpressible stink . . . and made me in like manner, ever and anon, belch, and as it were vomit back again at it.[15]

These were lively times for the geology of Italy. An early surviving diplomatic exchange, from the papal emissary Silvester Darius to King James v of Scotland in 1538, contains a reference to an eruption near Naples: 'Mount Vesuvius has set fire to the neighbouring countryside, consuming innumerable people and almost the town of Puteoli.'[16] Darius was wrong in saying that this was Vesuvius; it was in fact the eruption in the Phlegraeian Fields, west of Naples, that gave birth to Monte Nuovo. When Etna erupted in 1669, lava flowed as far as Catania seventeen kilometres away, surrounding and partially demolishing its walls. In the first recorded attempt to control the course of a lava flow, citizens of Catania attempted to break open the solidifying side walls of the lava as it pressed against the city to turn it sideways. This directed the flow to the nearby village of Paterno, whose residents fought a pitched battle with the Catanians to stop their engineering plans.[17]

Eruption of Mount Etna in 1766, in a late 19th-century engraving.

The English diarist and traveller John Evelyn (1620–1706) made a journey across the Alps and down into Italy in the mid-1640s. Just over ten years before, the area around Naples had been devastated by the most powerful eruption since AD 79. Evelyn described what he had heard of the 1631 eruption and the tsunami that followed it:

> It burst out beyond what it had ever done in the memory of history; throwing out huge stones and fiery pumices in such quantity as not only environed the whole mountain, but totally buried and overwhelmed divers towns and their inhabitants, scattering the ashes more than a hundred miles, and utterly devastating all those vineyards where formerly grew the most incomparable Greco; when, bursting through the bowels of the earth, it absorbed the very sea, and, with its whirling waters, drew in divers galleys and other vessels to their destruction, as is faithfully recorded.[18]

But now all was quiet, and Evelyn enjoyed the view:

> We at the last gained the summit of an excessive altitude. Turning our faces towards Naples, it presents one of the goodliest prospects in the world; all the Baiae, Cuma, Elysian Fields, Capri, Ischia, Prochita, Misenus, Puteoli, that goodly city, with a great portion of the Tyrrhenian Sea, offering themselves to your view at once, and at so agreeable a distance, as nothing can be more delightful.

John Evelyn, *Naples, from Mount Vesuvius*, 1645, pencil and chalk.

Climbing the last few hundred feet, however, Evelyn saw that this was no ordinary mountain:

> The mountain consists of a double top, the one pointed very sharp, and commonly appearing above any clouds, the other blunt. Here as we approached, we met many large gaping clefts and chasms, out of which issued such sulphureous blasts and smoke, that we durst not stand long near them. Having gained the very summit, I laid myself down to look over and into that most frightful and terrible vorago, a stupendous pit of near three miles in circuit and half a mile in depth, by a perpendicular hollow cliff (like that from the highest part of Dover Castle), with now and then a craggy prominency jetting out. The area at the bottom is plain like an even floor, which seems to be made by the wind circling the ashes by its eddy blasts.

After intensive examining and contemplating of volcanoes by travellers across the seventeenth and early eighteenth centuries, it comes as something of a surprise to read that Dr Johnson's definition of 'Volcano' in his *Dictionary*, first published in 1755, is merely 'A burning mountain'. The archaic ring of these words, picking up as they do from Kircher's 'Burning and Fire-Vomiting Mountains', already nearly a century old, reflects nothing of the scientific progress of the preceding hundred years, and the developing understanding that volcanoes were very much more than mountains that appeared to burn. If not from Kircher, Johnson took his definition from philosophical writings such as those of the seventeenth-century scholar Thomas Browne, whom he quotes in his literary illustrations of the word: 'Navigators tell us there is a burning mountain in an island and many volcanos and fiery hills.' Johnson goes on to cite another seventeenth-century writer, Sir Samuel Garth, who used Homer as his source:

> When the Cyclopes oe'r their anvils sweat,
> From the volcanos gross eruptions rise,
> And curling sheets of smoke obscure the skies.

Only a sermon of the Revd Richard Bentley takes Johnson to even the mildest scientific explanation, that 'Subterraneous minerals ferment and cause earthquakes, and cause furious eruptions of volcanos, and tumble down broken rocks.'

We need not expect Samuel Johnson to engage embryonic science in his definitions; after all, his entry for 'Chemistry' directs the reader to 'Chymistry' where his definition gives us:

> derived by some from χυμώ, juice, or χύω, to melt; by others from an oriental word, kema, black. According to the supposed etymology, it is written with y or e.

That is as far as he goes on the subject. A 'chymist', according to Johnson, is 'a professor of chymistry; a philosopher by fire'. So in the mid-eighteenth century, with the pioneering natural philosophers Robert Boyle and Robert Hooke well behind him, Johnson was as non-committal as it is possible to be.

Kircher's treatise *The Vulcanos* was followed by the 1743 English translation of *The Natural History of Mount Vesuvius* by the Neapolitan physician Francesco Serao (1702–1783). This gave a full account of the processes by which Vesuvius changed its shape from eruption to eruption, a mutation that would not come to be fully described until Sir William Hamilton published his *Campi Phlegraei*, with gouaches by Peter Fabris, 33 years later. Serao writes:

> The eruption of 1730 deserves our notice, not on account of its fury, but because it made a sensible alteration of the summit of the Volcano, for a great quantity of combustible and liquid matter, settling near the mouth of the Volcano, rendered the top much higher and more pointed than it was before.[19]

Serao's close observation of the eruption continues with the words:

> Another particularity remarkable in the same eruption was, that the flames were much brighter and livelier than usual, and rose into the air to a prodigious height. The fiery torrent

Anonymous, *Torre del Greco Devastated by the Eruption of Vesuvius*, 1798, gouache.

> which descended on the slope of the mountain, made no great progress, but on the side, where the southern borders of the Volcano, were shelter'd by the rocky circuit of Mount Somma, a horrid inundation of fire covered all the bottom of the plain, which we call the valley of Atria.[20]

Following the example of the people of Catania, when Etna erupted in 1669, the local villagers took direct physical action against the volcano:

> The chief damage sustained arose from the burning cinders, that set fire, to a large wood in the district of Ottaiano, which would have been entirely consumed, if, by cutting down the trees that lay in the way, a stop had not been put to the progress of the flames.[21]

Such eruptions were shocking and life-changing, and during the eighteenth century it must have appeared to residents that they would be tormented by them forever. One shaken English eyewitness reported of the July 1737 eruption:

> One can scarce frame to oneself a sight of greater destruction; ten successive Northern winters could not have left it in a worse condition, not a leaf or a tree, vine or hedge to be seen all the way we went . . . Here and at the town they had a new Earth, about two feet deep, some said more, and by the account of the miserable Inhabitants who were a dismal spectacle (tho' they had recovered their fright) they had a new Heavens for 18 hours but very different from those which St Peter mentions . . . In one convent two or three Nuns were overlaid, a death which came far out of its way to those poor unhappy wretches, who had locked themselves up for mankind. At Somma on the North East side it has made great havock, a Monastery of Nuns was destroy'd, and the Drones soon fled & dispers'd 'tis hoped to make up the time they have lost.[22]

The typical response of Neapolitans in the eruptive periods of Vesuvius was not to run away, but to bring out the portrait of St Januarius (San Gennaro), a third-century bishop who had been martyred in the lava fields west of Naples. He became the patron saint of the city, and was believed to have a salutary effect on the lava flow and to stop it in its tracks.

British natural philosophers travelled south to Vesuvius and Etna and in 1772 north to Iceland to make their early studies of volcanoes. Patrick Brydone climbed Etna in 1774, and in a series of letters to William Beckford extolled the majesty of what he saw:

> The immense elevation from the surface of the earth, drawn as it were to a single point, without any neighbouring mountain for the senses and the imagination to rest upon; and recover from their astonishment in their way down to the world. This point or pinnacle, raised on the brink of a bottomless gulph, as old as the world, often discharging

Anonymous, *Christ with St Januarius*, 16th century, pen and ink.

rivers of fire, and throwing out burning rocks, with a noise that shakes the whole island.[23]

Other empire-building nations commissioned exploratory expeditions to volcanoes in their newfound lands. A Spanish expedition under Don Francisco Suero climbed the volcano Misti, near Arequipa, Peru, during a period of activity, and only a few weeks before it erupted with enormous violence in July 1784.[24] Suero's report illustrates the dangers of climbing a rumbling volcano, and describes the fumes already given off by the lava and the putrid water.[25]

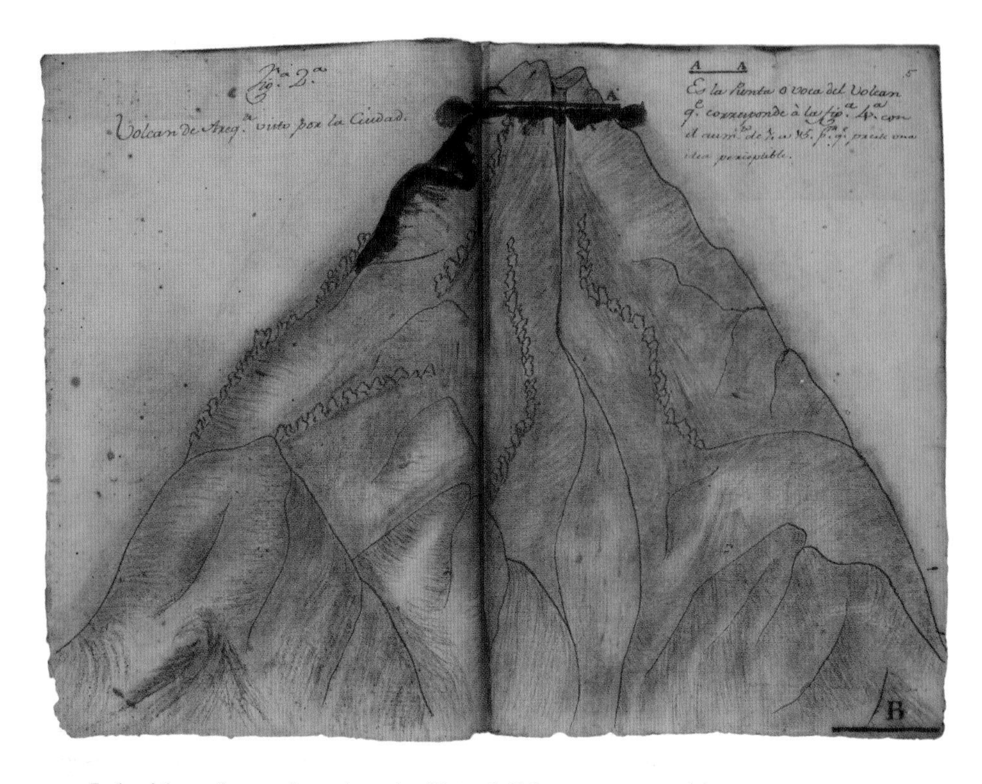

Anonymous, *Exploration of the Volcano Misti, near Arequipa, Peru*, 1784, pen, ink and watercolour on paper.

It had long been clear that the British Isles, incomparably rich though it is in geological variety, has had no active volcanoes for about 350 million years. Nevertheless, geologists refused on the face of it to take no for an answer. The itinerant lecturer and chemist John Warltire (*c.* 1725/6–1810) went so far as to observe that Exeter was built on an extinct volcano, and reported as much to the Chapter House Philosophical Society in London in 1785.[26] Further, Thomas Curtis of Bath wrote to Sir Charles Blagden, the secretary of the Royal Society, with an account of what seemed to be a new volcano in Flintshire, North Wales.[27] A generation earlier, a traveller in County Kerry, Ireland, believed he had discovered a volcano among cliffs by the Shannon estuary, and wrote to the Bishop of Kenmore to tell him so:

> Near two years ago a piece of one of these Cliffs fell off, whereupon there broke out a smoak attended with strong

> sulphureous smell . . . it has continued burning ever since . . . the Heat was so great, and the sulphureous stench so strong, that (though Pliny's Fate had not come into my mind) I could not wait to be over curious.
>
> From hence I descended obliquely to the Bottom of the Cliff where I had a full View of it, and of the Progress the Fire had made in it. It was entertaining to look up and discern the different Figures into which the Fire had eat it, and the Variety of beautiful Colours according to the different Minerals and Stones it met with . . . The whole face of the Cliff seemed to be a Composition of Red, yellow, black and white calcined stones and ashes of clay cemented together by Streams of melted Sulphur and Copperas that run among them like the Cement that Masons pour into Walls . . . I was informed by the inhabitants, that they see the Flame very plainly by Night, and I could observe the Air over it in a tremulous Motion like the Air over a burning Lime-kiln.[28]

Volcanoes were now coming rapidly up the scientific agenda. In a letter of 1781, otherwise about chemical retorts, Joseph Priestley (1733–1804) discussed the nature of lava with his friend Josiah Wedgwood (1730–1795):

> It is of some consequence to determine whether the lava come out of the volcano in the stonelike state in which we now find it, or acquires its present consistence and power of yielding air afterwards. I hope to find this, and I can have the use of a glass house fire for the purpose, of which I hope to make good use, as well as of my own.[29]

As a subject for study and observation, volcanoes were from the sixteenth to the eighteenth centuries led by science. It was not until 1726 that James Thomson, in his long poem 'Winter' from *The Seasons*, wrote of 'Hekla flaming through a waste of snow'. Following where scientists were beginning to explore were travelling artists. When Joseph Banks visited Iceland in

1772 his party included the artist John Cleveley the Younger (1747–1786), who brought back a series of workmanlike watercolours; when William Hamilton took the scientific initiative on Vesuvius in the 1760s he hired Peter Fabris, who revolutionized the art of the volcano, and changed our ways of seeing them.

4 Sir William Hamilton and the Lure of Vesuvius

In June 1801, Sir William Hamilton (1731–1803), a retired, 71-year-old distinguished diplomat who had represented British interests and witnessed and influenced triumph and disaster in the Kingdom of the Two Sicilies, wrote to Sir Joseph Banks offering a set of important diaries to the Royal Society. During his nearly 40-year period of service as British Envoy in Naples between 1764 and 1800, Hamilton had amassed one of the greatest collections of Greek and Roman pottery, other antiquities and books and manuscripts of immense value. (This is now split between the Royal Society, the British Museum and the British Library in London.) Demands upon Hamilton of a diplomatic nature were intermittent, giving him plenty of free time to explore archaeological sites and private and royal collections of art and antiquities, and above all to indulge his amateur interest in science. The diaries he offered to the Royal Society had been kept by the Italian priest and scientist Father Antonio Piaggio (1713–1797), who had recorded in words and drawings the day-to-day activities of Vesuvius between 1779 and 1794.

William Hamilton was, like Pliny the Younger, truly the right man at the right time in the right place, for it was during his tenure in Naples that Vesuvius entered one of its cyclical periods of intense volcanic activity. Sir William and his first wife Catherine lived quietly at Portici, with Vesuvius an active and entertaining feature in the view, as David Allan's touching portrait records. Using his own close observations of the volcano, and later with the help of Piaggio and others, Hamilton wrote a series

David Allan, *Sir William and Lady Hamilton at Home in Posillipo, Naples*, 1770, oil on copper.

of informed papers on the unfolding of the 1770s eruption cycle of Vesuvius, and sent these to the Royal Society. Piaggio, who lived at Resina at the foot of Vesuvius, 'always rose at daybreak', Hamilton wrote, 'and took his observations several times in the day. No man was ever more ready with his pencil as his masterly sketches testify nor no man was ever more attached to truth.'[1] Presenting Piaggio's eight bound manuscript volumes to the Royal Society, Hamilton confessed that he was 'too old & indolent to set about making an Extract that he thinks would form an interesting paper in some future volume of the Transactions of the Society.' But he added:

> Sir Wm is obliged to confess that they do not throw any certain lights into what is carried on by nature deep in the bowells of the Earth, but by the extraordinary & various forms that the smoke from the crater take in the air.

Hamilton had previously employed Peter Fabris (*fl.* 1756–84), an English-born artist then resident in Naples, to produce a set of gouaches, 59 of which were engraved and hand-coloured for publication in 1776. Being very much aware of the care required to capture the unique pictorial qualities of the Naples landscape, Hamilton employed Fabris 'to take drawings of every interesting spot, described in my letters, in which each stratum is represented in its proper colours'.[2] As a result, the paintings have an extraordinary fidelity to what was actually to be seen – the shocking beauty of the violent invading red and orange pyroclastic flow, blanketing with demonic carelessness the lower slopes of the mountain, and rendering its inhabitants either dead or miserable. It was of little immediate comfort to the locals to know that what was spewing out of Vesuvius now would make the slopes fertile and abundant again in a generation. Other paintings by Fabris show the successive crater shapes carved by the roaring emissions, the labours of clearing the lava and, once cooled, cut and polished, even the different colours and figures in specimens of the volcanic rock. William Hamilton's volumes discussed all the volcanoes in the Campi Phlegraei, 'Fields of Fire', as the landscape west of Naples is described, as well as the Vesuvius eruptions of 1767 and 1779, and those of Mount Etna. The church registers of Naples Cathedral, which recorded the

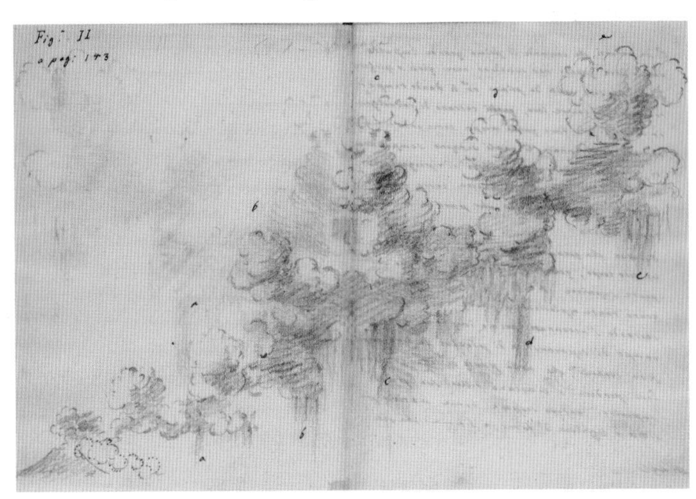

A page from Padre Antonio Piaggio's *Diary of the Activities of Vesuvius* (*c.* 1794).

William Hamilton and Pietro Fabris, plate from *Campi Phlegraei* (1776–9), depicting the eruption of Vesuvius, 23 December 1760 to 8 January 1761.

dates on which the saints' relics were carried to the foot of the mountain, gave him a fairly accurate picture of earlier eruptive activity. He looked also at the geology of the district, and at the condition and employment of the inhabitants.[3] In the introduction to the first volume Hamilton wrote:

> There is no doubt but that the neighbourhood of an active Volcano, must suffer from time to time the most dire calamities, the natural attendants of earthquakes, and eruptions. Whole cities, with their inhabitants, are either buried under showers of pumice stones and ashes, or overwhelmed by rivers of liquid fires; others again are swept off in an instant, by torrents of hot water issuing from the mouth of the same volcano, of which, in the very little we know of the History of Mount Vesuvius and Etna (the present active Volcanoes of these Kingdoms) there are many examples; and the ruins of Herculaneum, Pompeii, Stabia, and Catania relate their sad catastrophes in the most pathetick terms.[4]

The more active Vesuvius seemed to be, the keener Hamilton was to climb up it. He made many ascents of the crater, sometimes spending a night on the slopes:

William Hamilton and Pietro Fabris, plate from *Campi Phlegraei* (1776–9), showing specimens of volcanic rock from Solfatara, at Pozzuoli near Naples.

> I passed the whole day and night of the 12th [March 1765] upon the mountain and followed the course of the lava to its very source; it burnt out the side of the mountain within half a mile of the mouth of the volcano, like a torrent, attended by violent explosions . . . the adjacent ground quivering like the timbers of a water-mill; the heat of the lava was so great, as not to suffer me to approach nearer than within ten feet of the stream . . . large stones thrown onto it with all my force did not sink, but making a slight impression, floated on the surface, and were carried out of sight in a short time . . . with a rapidity equal to that of the river Severn, at the passage near Bristol.[5]

Another writer, whose identity is unknown, had an equally dramatic experience on the mountain as Hamilton, and in 1805 conveyed the sense of sheer excitement at the sight of the lava flow:

> Fancy to yourself an immense heap of burning coke which it resembles more than any other thing a mile in width and ten or twelve feet thick covering in its progress every thing that stood in its way. However seeing it did not advance so

> quickly as I expected I was tempted several times to go nearer in fact so near as I pulled out a piece of the Ashes which heated my Face not a little. To see it move on was like Magic, for trees, walls, houses, appeared to fall before it as soon as it touched it in the least and one would frequently see it stop and collect in a moment tho' the great mass was running on burning with trees as large as those in our church yard at Exeter floating upright enveloped by the Flames 'till they were absolutely melted away to nothing, such my dear Father is our Eruption, which in my opinion is one of the greatest sights that ever can be seen.[6]

The volcano was not only attractive as a subject for artists; it was also a useful step on the diplomatic round, and a source of entertainment. Hamilton accompanied King Ferdinand IV of Naples and his queen up Vesuvius during an eruption, and, glowing in the heat, together they watched a current of lava flow down a 15-metre (50-foot) cascade, and along a hollowed-out channel.[7] The French painter Joseph Franque (1774–1833) recorded another visit to the smoky summit by King Ferdinand, his son Prince Francis and the Duke of Calabria. Seated is Archduchess Maria Clementina of Austria, wife of Prince Francis. The party of nine are fragile and vulnerable in their military finery, with medals, ribbons and feathery hats, all so touchingly inappropriate in the bleak Goyaesque landscape.

Naples attracted foreign artists in the mid- and late eighteenth century, such as the Frenchman Pierre-Jacques Volaire (1729–1799), Joseph Wright of Derby (1734–1797) and the Austrian Michael Wutky (1739–1822). Naples was the capital of an independent nation, and a rich and prosperous trading port that kept its political distance from the Pope in Rome while balancing its economic and political relations with Britain, France and the Austro-Hungarian Empire. Enterprising machinations that such positioning demanded contributed to the growth of an economy in which an art market was an essential component, so it was not only Vesuvius that was the attraction, but the entire bubbling stewpot of supply and demand in full swing. Along with Volaire,

William Hamilton and Pietro Fabris, plate from *Campi Phlegraei* (1776–9), showing Hamilton accompanying the King and Queen of Naples up Vesuvius during an eruption, 11 May 1771.

artists who were supplying the market with paintings of the volcano at this period included Camillo de Vito, Xavier della Gatta and many others whose names have not survived.

Volaire studied with Vernet in Paris, and moved to Italy in 1764, settling in Naples on the lido at Chiaia, directly across the bay from the volcano.[8] There he became greatly sought after as a painter of the most dramatic Vesuvian eruptions, a speciality he prudently developed after the 1771 eruption, often making strong contrasts between lava-light and moonlight. His accomplished technique and bold manner influenced many artists around him, including Wright and Wutky, and he was a favourite with William Hamilton, who bought his work.[9] Volaire was a generation younger than the painters of the Rococo style, and moved Rococo subject-matter on from nymphs and shepherds in pastel greens and pinks to the fiery violence of volcanoes. He brought power and passion to this fey and often feckless style that had already become unfashionable in the face of developing revolutionary tendencies in Europe and the emergence of Neo-Classicism and the Romantic. Where true Rococo might use a series of artfully aligned corn sheaves, gardening tools or shells to create an interlinked vertical climbing form in a painting or piece of furniture, Volaire uses in *Vesuvius Erupting at Night* a

Joseph Franque, *Visit to Vesuvius by King Ferdinand of Naples and his Party,* 1815, oil on canvas.

Xavier della Gatta, *Eruption of Vesuvius*, 1794, gouache.

group of gesticulating figures and a shattered tree that dance in silhouette as a foil to the devastating explosion.

Pierre-Jacques Volaire, *Vesuvius Erupting at Night*, 1770s, oil on canvas.

Among the British artists who, like Wright of Derby, visited Naples in the mid- and late eighteenth century, were the Welshman Thomas Jones (1742–1803)[10] and, very briefly, the Scot Jacob More (1740–1793). Wright visited Naples for only about four weeks, from early October to early November 1774, and was for part of that time a guest of Sir William Hamilton. Four weeks in Naples furnished Wright with material if not for a life's work then certainly for a brave posthumous reputation. During his time there, Vesuvius performed for Wright magnificently, as he told his brother Richard in a letter from Rome: 'there was a very considerable Eruption at the time, of which I am going to make a picture – Tis the most wonderful sight in nature.'[11] The vivid gouache *Vesuvius in Eruption* is undoubtedly an immediate record of the eruption that Wright witnessed.[12]

Wright suffered from melancholy or, as he called it, 'torpor': 'I have now dragged over four months, without feeling a wish to

Joseph Wright of Derby, *Eruption of Vesuvius*, 1774, gouache.

take up my pencil', he wrote in 1783.[13] This condition recurred, and we might suppose that it had a deleterious effect on his production as an artist, to the extent that he retreated into his Italian subjects not just to make money from their sale but to provide certainty and a measure of consolation. Wright's Italian trip contributed other subjects to his repertoire, including Lake Nemi, Lake Albano and the Girandola, the annual firework display at the Castel di Sant'Angelo in Rome. His *View of Catania, with Mount Etna* has a limpid calm that the Vesuvius subjects characteristically lack, and may have been intended as a pendant for an erupting Vesuvius.[14]

While his paintings of Lake Nemi and Catania evoked calm and stability, feelings of anger, uncertainty and lack of confidence could all be sublimated by a colourful, violent eruption. The earliest of Wright's big eruption canvases, *Eruption of Vesuvius, seen from Portici*, William Hamilton's villa, may have been started when he returned from Naples to Rome in November

Joseph Wright of Derby, *View of Catania, with Mount Etna*, c. 1775, oil on canvas.

Joseph Wright of Derby, *Vesuvius in Eruption with a View over the Islands and Bay of Naples*, *c.* 1774–6, oil on canvas.

1775. It was bought by John Leigh Phillips, who already owned a number of Wright's paintings. The second version, *Vesuvius in Eruption with a View over the Islands in the Bay of Naples*, was painted a year or two later. A third canvas was exhibited in 1778 and bought immediately by Catherine the Great of Russia, while a fourth remained in the artist's collection and was offered for sale after his death. All reveal how Wright stalked the mountain from points around its base, showing glimpses of the Sorrento peninsula, the mole and lighthouse in Naples harbour, and the bay itself.

Thomas Jones had more luck with the timing of his visit to Naples. In his first sight of the city in September 1778 he saw 'Mount Vesuvius at a distance belching forth Volumes of Smoke, which formed a long streak of a Cloud as far as the Eye could reach.'[15] A day or two later he climbed the mountain, where he saw the lava which

Thomas Jones, *Mount Vesuvius from Torre dell' Annunziata near Naples*, 1783, oil on paper laid on canvas.

roll'd and tumbled and was composed of Melted Matter fluid and glowing underneath, with large black irregular blocks of Scoriæ floating on the Surface. Just below this new Stream we crossed some old Lava & came to a Vineyard the greatest part of which had been destroy'd by the present Eruption but the day before together with the house belonging to it . . . It was curious to observe the Bunches of grapes shrivel up like raisons and the leaves wither & take fire upon the approach of this tremendous burning River.[16]

Jones recorded in his memoirs how this eruption brought other Britons down from Rome at high speed, one of the party being Jacob More, who made a number of what he called 'flying sketches'. As Jones recalled, 'none of the Company waited a moment for him, he contrived to keep up with the party and brought back a dozen *Views*, and these were to pass as *portraits* of the respective scenes.'[17] While More's *Eruption of Vesuvius* (1790s) includes the diminutive figures of Pliny the Elder and his companions facing death, its subtitle, 'The Last Days of Pompeii', is a mid-nineteenth-century addition following the publication in 1834 of Edward Bulwer Lytton's successful novel.

Perhaps unconvincingly, Wright of Derby wove a volcano subject into a painting in subsequent years. His *Widow of an Indian Chief Watching the Arms of her Deceased Husband* (1785) shows in the distance a fanciful American volcano smoking and glowing while thunder and lightning rage beside it. The mourning foreground figure is carrying out the custom imposed on widows of Native American chiefs, to sit every day under a totemic tree bearing his bow and arrows and tomahawk for the full 28-day cycle of the first moon after the husband's death. As Wright put it in the paragraph accompanying the painting's exhibition in Covent Garden in 1785, 'She remains in this situation without shelter, and perseveres in her mournful duty at the hazard of her own life from the inclemencies of the weather.'[18] Wright here uses the erupting volcano as a sign of the most terrible weather that he could possibly imagine the widow could suffer. It is an early example of a volcanic metaphor being used with serious artistic intent.

Joseph Wright of Derby, *The Widow of an Indian Chief Watching the Arms of her Deceased Husband*, 1785, oil on canvas.

While artists expressed an immediate sense of wonder at the violence and beauty of volcanoes, scientists climbed up them to take measurements. The young Michael Faraday (1791–1867) wrote excitedly about climbing Vesuvius with his teacher Sir Humphry Davy in May 1814. At the summit Faraday found that

> the volume of smoke and flame appeared immense, and the scene was fearfully grand . . . at times we could see the flames breaking out from a large orifice with extraordinary force, and the smoke and vapour ascending in enormous clouds; and when silence was made the roaring of the flames came fearfully over the ear . . . I incautiously remained to collect some of the substances, and was then obliged to run over the lava, to the great danger of my legs.[19]

On a return visit in March 1815 Faraday was even more reckless as the mountain shivered and shook:

> I heard the roar of the fire and at moments felt the agitation and shakings of the mountain but . . . [we] went forward and we descended some rocks of lava and proceeded onwards towards the very edge of the crater leaping from one point to another being careful not to slip not only to avoid the general inconveniences of a fall but the being burnt also for at the bottom of the cavity the heat was in general very great . . . The ground was in continual motion and the explosions were continual . . . then might be seen rising high in the air numbers of redhot stones and pieces of lava which at times came so near as to threaten us with a blow.[20]

The courage of these early scientists was extraordinary and unnerving. So cool were they that this party fried eggs on a piece of lava, ate a hearty lunch and sang 'God Save the King' as earthquakes made the mountain shake like jelly.

The scientist and writer Mary Somerville (1780–1872) got as close to Vesuvius in fact and in artistic expression as any artist of her generation. She had been lulled into a sense of false security before she and her family party left for the mountain, telling her mother-in-law that it 'does not even smoke at present so we shall get to the top – it is not very high and is said not to be a difficult undertaking'.[21] With her daughter Margaret riding on the shoulders of a guide, Mary and her family climbed the mountain in March 1818:

> At first we went through deep ashes sliding back more than half at each step then we got a stream of hard lava like the steepest stair or rather ladder, however perseverance overcame everything by going slow, and stopping after we came to the lava which had flowed in the irruption which took place three months ago. It was not smoking and in many places red hot. We sat down and having roasted our eggs in it we dined it was like sitting on a furnace – after our meal we began the ascent to the crater. This was still more perpendicular, and so hot that our feet were roasted and sticking paper in the crevices it flam'd ... Now the fumes of sulphur became oppressive but we continued our ascent when all at once we came to the great Crater. My feelings were so strong of astonishment and still more of terror that I stood motionless without uttering a word. The extent is enormous, the depth unknown, the inside rough shaggy horrible, smoking fiercely from every crevice, brilliant with all colours, deep black bright red green yellow orange which all formed by the vapour, the heat and the smoke was distressing but we were determined to go round it which took us about an hour, at one place we stuffed handkerchiefs into our mouths and noses and ran as fast as we could over a place that was so thick of vapour we could hardly see one another the red flame issued from many places and the heat so great that I several times thought my petticoats were on fire all this while we were going round the verge of the Crater, at one place there were large deep holes from which steam rushed out violently ... With joy we began the descent at a place where there was no lava and we ran as hard as it was possible literally up to the knees in ashes at every step.

> We were at the bottom of the cone in half an hour, mounted our asses returned to Resina where the carriage waited us and got back to Naples by five o'clock tired indeed but not too much so but dirty beyond imagination.[22]

Mary Somerville, through courage and a powerful sense of adventure, saw the grumbling mountain from its very lip, as did Humphry Davy and Michael Faraday. All reported the intense colours and heat of the active volcano. The historian Anna Jameson, on the other hand, noticed the noise, taking a longer view in 1826:

> Mount Vesuvius is at this moment blazing like a huge furnace; throwing up every minute, or half minute, columns of fire and red-hot stones, which fall in showers and bound down the side of the mountain. On the east there are two distinct streams of lava descending, which glow with almost a white heat, and every burst of flame is accompanied by a sound resembling cannon at a distance.[23]

Volcanoes were not, however, just magnificent natural firework displays for those lucky enough to be able to watch from a safe distance. They had knock-on effects that by the late eighteenth century were becoming closely observed in Britain and Europe. Distant volcanoes were gradually seen to be the cause of new erratic behavioural patterns in the weather. The volcanic fissure running northeast across Iceland cracked in 1783, and for eight months, from June until the following February, spewed out lava and sulphur dioxide. This reacted with the water in the atmosphere to create a poison belt of sulphuric acid. The volumes of lava were so great along a line 30 or 40 kilometres (around 20 or more miles) long, and the poisons so overwhelming, that 9,000 square kilometres (3,500 square miles) of land were covered by lava or ash, and crops and cattle perished.[24] Within weeks the contagion had spread south. Its effects were observed by the naturalist Gilbert White (1720–1793), who described events in his garden at Selborne in Hampshire:

> The summer of the year 1783 was an amazing and portentous one, and full of horrible phenomena; for, besides the alarming meteors and tremendous thunderstorms that affrighted and distressed the different counties of this kingdom, the peculiar haze, or smoky fog that prevailed for several weeks in this island, and in every part of Europe, and even beyond its limits, was a most extraordinary appearance, unlike anything known in the memory of man. By my journal I find I had noticed this strange occurrence from June 23rd to July 20th inclusive, during which period the wind varied to every quarter without any alteration in the air. The sun, at noon, looked as blank as a clouded moon, and shed a rust-coloured, ferruginous light on the ground, and floors of rooms; but was particularly lurid and blood-coloured at rising and setting. All the time the heat was so intense that butchers' meat could hardly be eaten on the day after it was killed; and the flies swarmed so in the lanes and hedges that they rendered the horse half frantic, and riding irksome. The country people began to look with a superstitious awe at the red, lowering aspect of the sun; and indeed there was reason for the most enlightened person to be apprehensive; for, all the while, Calabria and part of the Isle of Sicily were torn and convulsed with earthquakes; and about that juncture a volcano sprung out of the sea on the coast of Norway.[25]

The footnote to this letter in the 1941 edition of *The Natural History of Selborne*, edited by the naturalist James Fisher, innocently states that the 'peculiar haze' and all that followed was 'probably due to a great volcanic explosion far away'.

Another portentous year was 1816. In April 1815, when Europe was preoccupied with political and military cataclysms of its own, Mount Tambora exploded on the island of Sumbawa, east of Java. This distant volcano on the other side of the world killed 10,000 people almost immediately, destroyed the Tambora and Sanggar kingdoms, eradicated the Tambora language and brought famine and disease in its wake. Once the local after-effects had run their course, at least 117,000 people had died.[26]

Devastation far away caused, first of all, extraordinarily vivid sunsets in Europe. Luxurious light effects were soon followed by months of appalling weather: it poured with rain all night before the battle of Waterloo, 18 June 1815, creating appalling muddy conditions to add to the chaos and horror of the day. Lord Byron used the state of the weather to emotive effect in verses in *Childe Harold's Pilgrimage*:

> The thunder-clouds close o'er it, which when rent
> The earth is covered thick with other clay,
> Which her own clay shall cover, heaped and pent,
> Rider and horse, – friend and foe, – in one red burial blent.[27]

It rained all the next summer too: 'Rain Rain Rain', the painter J.M.W. Turner complained when he was stuck with his horse for weeks in rutted roads in Yorkshire.[28]

The decades around the turn of the nineteenth century were ones of very high volcanic activity worldwide, with eruptions in Italy, Iceland, Java and, in 1812, on the island of St Vincent in the Windward islands of the West Indies. It is significant that this intense geological activity coincided with decades of revolution in Europe; indeed, the two phenomena, geological and political, have many distinctive links. The Laki eruption of 1783 devastated crops all over Europe, which raised the price of bread, one of the triggers of the French Revolution six years later. Never before had scientific and literary observations on volcanic activity been so comprehensively and effectively catalogued. Immediate responses scribbled or painted in the heat of the moment led to more measured scientific writing, based on data collected on the spot and then considered far away, at leisure. The naval officer Thomas Cayley witnessed the eruption on 30 April 1812 of Soufrière, a volcano on St Vincent. He described himself to his sister, in a letter written four months after the blast, as being 'confused and scattered by the horrors of the ever memorable eruption'. Such is the terror and violence of serious eruptions that all who write about them in this period seem to suggest that the event that they have witnessed is worse than any other. Themes recur in these writings: comparison with the noise of a military cannonade; the simile of the river; the depth of horror; the red-and-black shagginess of the volcano's interior. They really do still, in the eighteenth century, feel that they are looking into the depths of hell: a shadow of the Middle Ages, of 'primitive' legend and myth, has not quite gone away. Personal experience blankets out relativity and masks any sense of proportion:

> The various accounts of it, however, which I discern inscribed in the newspapers, will convey but a very faint idea of the sublime horror of that ever memorable night. History perhaps does not record a more dreadful convulsion for I believe excepting the eruptions of Mount Hekla in Iceland, those from Vesuvius and Etna are not to be compared to it either in point of duration or volume.
>
> The ashes fell upon a ship's deck 600 miles to the Eastward of Barbados, which is at least 60 miles distant from this Island. The repeated explosions which continued all night and till 6 o'clock the following morning, can only be compared to the roaring of ten thousand cannon, for the reports were heard distinctly in lee of the Windward and Leeward Islands continued to cover many of them during the greater part of the day following . . . During the eruption a new crater was formed . . . the old crater is in circumference about three miles, and one in diameter, approaching in form very near a circle. Its depth appears to be about 1,800 feet.[29]

Academic and political responses followed in a tide after the echoes of the event itself had died down. The eruption of Souffrière sparked a meeting of a House of Commons committee, and soon after a discussion of the event was organized by the Geological Society.[30] J.M.W. Turner exhibited a painting of the Souffrière eruption at the Royal Academy in 1815, by coincidence hanging at the time of the Tambora eruption, whose after-effects were to cause him such trouble in Yorkshire the following year. Turner' s painting, whose exhibited title was *The Eruption of the Souffrier Mountains, in the Island of St Vincent, at Midnight, on the 30th of April, 1812, from a Sketch Taken at the Time by Hugh P. Keane, Esqre*, was made not from his own observation but from a sketch, so far untraced, which was drawn by the barrister and sugar plantation owner Hugh Perry Keane. Turner was, as the title indicates, at pains to show that the painting was not an eyewitness account, but he nevertheless instilled an extraordinary sense of drama into this night scene seared by the blazing light and violence of the eruption. There is no other trace of Keane in Turner's surviving correspondence or literature, so it is not clear how they met. It has however come to light that Turner had financial interests in West Indian slavery in the 1810s and early 1820s, and it is possible that this common interest may have brought the two men together, however briefly.[31] While Keane's drawing does not survive, his diaries do, and in them he gives a brief but vivid real-time account of what he saw:

> Wed 29 April 1812: . . . On to see Souffrier, involved in dark clouds, and vomiting black smoke. Thurs 30: . . . in the afternoon the roaring of the mountain increased and at 7 o'clock the Flames burst forth, and the dreadful Eruption began. All night watching it – between 2 and 5 o'clock in the morning, showers of Stones and Earthquakes threatened our immediate Destruction. Fri 1 May: . . . The whole Island involved in gloom ... The mountain was quiet all night . . . Sun 3: . . . a strange and dismal sight, the River dried up, & the Land covered with Cinders and Sulphur . . . burnt carcasses of cattle lying everywhere . . . Wed 6: . . . The Volcano again blazed away from 7 till ½ past 8. Thurs 7: Rose at 7. Drawing the eruption.[32]

This was probably the drawing that came Turner's way, and with it and perhaps conversation with Keane as inspiration Turner drew his own vivid word picture in a poem that shakes under its own powerful internal energy:

> Then in stupendous horror grew
> The red volcano to the view

J.M.W. Turner, *The Eruption of the Souffrier Mountains, in the Island of St Vincent, at Midnight, on the 30th April 1812 from a Sketch Taken at the Time by Hugh P. Keane, Esqre*, 1812, oil on canvas.

And shook in thunders of its own,
While the blaz'd hill in lightnings shone,
Scattering their arrows round.
As down its sides of liquid flame
The devastating cataract came,
With melting rocks, and crackling woods,
And mingled roar of boiling floods,
And roll'd along the ground.

This poem, probably written before he painted the picture to raise the intellectual steam he required, was published by Turner in the catalogue of the 1815 Royal Academy exhibition, and was intended thus to stand in the mind of viewers when they contemplated the painting. Interestingly, if one tries to imagine this painting as a day scene, and in one's mind's eye douses the fire and removes the lava bombs and exotica, this painting of a West

Indian volcanic eruption has all the hallmarks of a work in oils painted in the mountainous English Lake District or the Highlands of Scotland.

Trusted for decades for his truthfulness in landscape depiction, even Turner would make things up. We find within his *oeuvre* views in the West Indies, India, the Middle East and even a minuscule view of the Andes from the Chilean coast, despite the fact that he travelled to none of these places. There are also paintings of eruptions of Vesuvius, none of which he witnessed either. While he saw Vesuvius smoking, and even appears to have climbed to the crater,[33] the volcano did not do him the courtesy of erupting during the two weeks he was in Naples in 1819. What Turner did notice, however, and effortlessly depicted, was that in October 1819 Vesuvius was smoking from two points within the crater, and making two distinct plumes of white smoke which, before the wind combined them into one, were visible by his sharp eyes from as far away as the slopes above the city.

Turner's magnificent *Bay of Naples (Vesuvius Angry)* and *Eruption of Vesuvius* are fabrications, and we are dependent on artists of lesser talent for the first shocking eyewitness accounts

J.M.W. Turner, *Naples: Vesuvius, from Naples: Rome*, from his *Colour Studies Sketchbook*, 1819, pencil and watercolour on paper.

J.M.W. Turner, *Vesuvius in Eruption*, 1817, gouache.

of eruption. When Captain Tillard, anchored off San Miguel in the Azores, heard reports in June 1811 of an island rising out of the sea off Ponta da Ferraria, he immediately set sail in his frigate HMS *Sabrina* to see it. His artist made a dramatic study of the fountains of smoke and lava and, like Graham Island in the Mediterranean twenty years later, this scrap of hot, dry land was claimed for the British crown, and named Sabrina. It then vanished forever.[34]

Growth in knowledge of volcanoes led to growth in understanding and misunderstanding of them. Lord Byron told his publisher, John Murray, of an encounter in Ravenna between Humphry Davy and Lord Byron's mistress at the time, Teresa, Contessa Guiccioli:

> who by way of expressing her learning in the presence of the great Chemist then describing his fourteen ascensions of Mount Vesuvius – [Teresa Guiccioli] asked 'if there was

> not a similar Volcano in Ireland?' – My only notion of an Irish Volcano consisted of the Lake of Killarney which I naturally conceived her to mean – but on second thoughts I divined that she alluded to Iceland & to Hecla – and so it proved – though she sustained her volcanic topography for some time with all the amiable pertinacity of 'the Feminie'.[35]

John Martin (1789–1854), a difficult and driven artist who had travelled from Newcastle upon Tyne in 1806 to make his fortune in London, had by the early 1820s become the talk of the town.[36] In 1822 he showed a group of paintings at the fashionable Egyptian Hall in Piccadilly, including the *Destruction of Pompeii and Herculaneum*, which caused a sensation on its first exhibition. The young architect Ambrose Poynter told his friend Robert Finch, then resident in Rome, all about it:

> [Martin] has just finished another large picture – the destruction of Herculaneum and Pompeii of which the sky is loaded with chrome and vermillion, and the foreground, where he has attempted the same sort of effect as in Belshazzar, a complete manqué – But the content of landscape, comprising a bird's eye view over all the cities at the base of Vesuvius – Herculaneum – Oplentis – Pompeii – Stabia etc is wonderful – Admirably composed – with great attention to truth, and detailed in a most extraordinary manner.[37]

Martin made a good living and a controversial name for himself out of paintings and prints of cataclysmic death and destruction. However, through a quirky fault-line in his personality he came to squander it all in the 1830s and '40s by spending ten years of his life, and all of his fortune, on attempting and failing to establish a civil engineering infrastructure to enable London to have a freely running and permanent clean water supply. This was personal civic duty taken to the ultimate degree, and in undertaking it Martin became a figure of fun. In the days of his ascendancy, however, he was more popular than Turner: his paintings sold for many hundreds of pounds, and his engravings

John Martin, *Destruction of Pompeii and Herculaneum*, *c.* 1821, oil on canvas.

sold in their hundreds. He put immense pressure on himself to succeed, and as a young artist sent the narrative painting *Sadak in Search of the Waters of Oblivion* to the 1812 Royal Academy exhibition. 'To my inexpressible delight', he wrote afterwards, the work was written about in the newspapers, and sold. The subject derived from a then popular story, 'Sadak and Kalasrade', written in 1764 in the 'Persian' style by the English author James Ridley. The diminutive figure of the warrior Sadak climbs through a fiery underground river inside a volcano on a trial to win his love:

> He marched onward, the hot soil scorching his feet, and the sulphureous stenches blasting his lungs, till he perceived a huge cave, through which ran a rivulet of black water.[38]

Already the painting has what were to become the full 'Martinian' ingredients: the heated colour, the desperate circumstances, over-powering and terrifying natural forces. The trick Martin had successfully performed was to put fashionable rural landscape

John Martin, *Sadak in Search of the Waters of Oblivion*, 1812, oil on canvas.

subjects aside, and to set his sights on horror, excess and perversity. Here Sadak is in the very heart of a volcano, in sizzling heat, of a kind that contemporary travellers had tried to describe.

While Vesuvius and Etna were familiar to artists and the picture-buying public, volcanoes as distant as Hekla waited patiently to be discovered. Central Iceland was remote until the twentieth century even to Icelanders themselves, living as they still do overwhelmingly along the coasts and in particular in Reykjavík on the southwest promontory. Uno von Troil observed something that is usually common to all volcanoes:

> It scarcely ever happens that the mountains begin to throw out fire unexpectedly; for besides a loud rumbling noise, which is heard at a considerable distance . . . and a roaring and a cracking in the part from whence the fire is going to burst forth, many fiery meteors are observed.[39]

Von Troil also described some phenomena particular to Iceland, eruptions known as 'jokuls' which take place beneath glaciers, causing 'the bursting of the mass of ice with a dreadful noise . . . Flames then burst forth, and lightning and balls of fire issue with the smoke, which are seen several miles off.'[40]

Using considerable imagination, and perhaps with the help of travellers' tales, an unknown eighteenth-century Danish artist, sent to Iceland on the command of King Christian VI (*reg.* 1730–46) of Denmark, painted a series of dramatic canvases of Icelandic volcanoes erupting. He may not have experienced the real thing. The tall red protuberance, a monstrous and vulgar carbuncle that rises up out of the ground, differs markedly from a typical volcanic form, but is nevertheless as violently repulsive in its own way as a real eruption might be. Other paintings in the series are caricatures of rocky islands that have risen from the sea, or been thrown up inland, painted in a manner that owes more to Italian and northern Renaissance representations than to actual experience. These works are no closer to the actual event than is the fanciful view in the Krafft Collection of Hekla erupting: here the artist leads the viewer to believe that pleasure craft plied a lake

Anonymous, *Eruption of Icelandic Volcano*, early 18th century, oil on canvas.

Anonymous, *Eruption of Icelandic Volcano*, early 18th century, oil on canvas.

Anonymous, *Eruption of Hekla, Iceland*, late 18th century, gouache.

(non-existent) surrounding the volcano, and that a tidy viewing platform had been set up.

Volcanoes, with their extraordinary physical power and as yet not fully understood dynamics, were ripe settings for satire, as the artists of political prints had long discovered. For comic or satirical intent the metaphor of the volcano became common. James Gillray used it in 1794 to depict the uncontrollable terrors of the French Revolution. During that year alone over 15,000 people were guillotined in France. Gillray shows the head of Charles James Fox, the radical English politician, being carried, like the head of St Januarius, in procession around the Vesuvius as if it were the only charm that could possibly staunch the mountain. The bearers are Tory politicians, dressed as revolutionary *sans-culottes*, who are using their political foe as a talisman to stop the political eruption in France that threatens to destroy London, Vienna, Berlin, Rome, Flanders and Holland – all depicted symbolically in the engraving.[41]

James Gillray, *The Eruption of the Mountain; or, the Horrors of the 'Bocca del Inferno', with the Head of the Protector Saint Januaris Carried in Procession by Cardinal Archeveque of the Lazaroni*, 1794, hand-coloured etching and aquatint.

Gillray's *A Cognoscenti Contemplating ye Beauties of ye Antique* was published in 1801, the year after Sir William Hamilton came home, and presented a sad, reclusive image of a decrepit old man who had been cruelly cuckolded by Nelson and lost a beautiful wife. Here he is, bent and disappointed, looking through reversed spectacles at a marble bust of 'Lais', the very image of his wife with a broken nose and mouth. A bacchante holds one of Lady Hamilton's 'Attitudes' nearby, and behind is an Egyptian bull, a Cupid with a broken bow and arrow, a hare (the symbol of lechery), and various other broken antiquarian bits and pieces. On the wall are pictures of Cleopatra (Lady Hamilton with naked breast and a bottle of gin), Mark Anthony (Nelson), Claudius (Sir William Hamilton, miserably looking the other way) and, erupting in a great red spurt, Vesuvius.[42] Another satirical print, by Frederick George Byron, shows Edmund Burke, with volcanic words spewing from his mouth, fulminating against Fox and Sheridan: 'Black as ten furies! Jacobite miscreants . . . Terrible as Hell! Infernal spawn! . . . Pimps, Panders, Parasites, Devils'. The current event that prompted the cartoon

James Gillray, *A Cognocenti Contemplating ye Beauties of ye Antique*, 1801, hand-coloured etching.

was Burke's speech in the House of Commons denouncing the French Revolutionary Constitution, in which he quoted Milton's *Paradise Lost*: 'Black it stood as night, Fierce as ten furies, terrible as Hell'.[43] It is self-evident that these volcanic metaphors emerged in the public prints during the course of Vesuvius' late-eighteenth-century eruptive period, when the meaning of the cartoons would be all too obvious.

In exactly the same way as Vesuvius provided a metaphor for eighteenth-century cartoonists, so their twenty-first-century counterparts used the April 2010 eruption of Eyjafjallajökull to comment on the British General Election campaign, then underway. In *Eruption Smothers Britain*, Christian Adams shows the country smothered by ash-cloud portraits of the three party leaders, Nick Clegg, Gordon Brown and David Cameron. Gerald Scarfe, in *One Still Flying*, depicts Cameron and Brown trying to shoot down with bows and arrows a buoyant Nick Clegg, the Liberal Democrat leader. In the background a volcano, 'Public Unrest', is spewing forth a cloud of black smoke.

Rudolphe Raspe wrote in his comic *The Surprising Travels and Adventures of Baron Munchausen* (1785) how he was inspired by Patrick Brydone's account of his *Tour Through Sicily to Malta* (1780) to induce the eponymous baron to visit Mount Etna. Moving from the rational tone of Brydone to the manic voice of Munchausen, Raspe leaps, like Empedocles, into the crater of Etna. There he meets Vulcan and his Cyclopes, who entertained him liberally and healed the wounds and the burns he sustained in his fall. Vulcan explained how he threw red-hot coals at his assistants when he got angry, and they would parry them and throw them up out of the crater causing 'what I find you mortals

Frederick G. Byron, *The Volcano of Opposition*, 1791, hand-coloured engraving. Edmund Burke is in Parliament's House of Commons in London denouncing the French Revolutionary Constitution.

Gerald Scarfe, *One Still Flying*, April 2010. A political cartoon published in the *Sunday Times* on 4 April 2010, when the eruption of Eyjafjallajökull in Iceland had grounded most commercial flights. The Liberal Democrats led by Nick Clegg were then doing well in the opinion polls. Clegg is astride the Lib-Dem logo.

call eruptions'. Munchausen was then introduced to Venus, who showed him 'every indulgence which my situation required', lucky fellow. But soon becoming jealous, Munchausen's luck turned when Vulcan picked him up and dropped him into a well, from which he emerged in the South Sea Islands on the other side of the globe.[44]

5 The First Days of Graham Island and the Last Days of Pompeii

Some time towards the end of June 1831, sailors passing between Sciacca in southern Sicily and the island of Pantelleria, 120 km (74 miles) towards Tunis, became gradually aware of unusual turbulence in the water. The sea appeared to convulse, and shocks were felt on the hulls of boats. This was not uncommon: such convulsions had been reported before by sailors in volcanic areas. Captain Swinburne of HMS *Rapid* noticed it on 28 June, and over the next two or three days the shocks became more frequent. Local fishermen reported the sea becoming muddy, and bubbling. Their first optimistic thoughts were that a huge shoal of fish had appeared, but very soon hundreds of dead fish floated to the surface, and the air that bubbled up gave off a sulphurous stink.[1] Over the course of the following days, Sciacca suffered earthquakes; there was thunder and lightning; silver spoons turned rapidly black in the acidic air. A column of smoke, which many assumed came from a steamboat sailing to Malta, was visible far out at sea. The sea around the emerging island – for that is what it was – soon became alive with boats from Sicily, Naples, Sardinia and Malta, as well as from Britain, France and Spain, marvelling at this new manifestation of the power, unpredictability and opportunity of nature.

The British vessels HMS *Briton* and HMS *Rapid*, sent from Malta by Vice Admiral Henry Hotham, were joined on 17 July by the brigs HMS *Adelaide* and HMS *Philomel* and the cutter HMS *Hind*. The despatch of five British naval ships suggests intense political interest at the very least. By now, the island had grown

British School, *Graham Island*, 1831, watercolour.

to between 20 and 25 metres (70 and 80 feet) in circumference, and had become the sea-washed platform for an explosive display of fountains of fire 180 metres (600 feet) high.

The island grew and grew until by 22 July it was a 1.2 km (three-quarters of a mile) circuit, and 25 metres (eighty feet) high on its northwest side. Following the naval officers, armed with telescopes, sextants, plumb lines and flags, came the scientists with notebooks, barometers and bottles: Carlo Gemmellaro from Sicily, Friedrich Hoffmann from Germany, Constant Prévost from France, and for Britain Dr John Davy, the head of the Medical Corps in Malta, and younger brother of Sir Humphry Davy. They measured and noted and sniffed at the sulphurous air: Prévost asserted that the event was like uncorking a bottle of champagne, while Davy likened it to pistol shots and musket fire.[2] Prévost named it 'Ile Julie' because it had arisen in July; Captain Senhouse of the *Hind*, planting the Union Jack, considered the eruption to be a 'permanent island' on 2 August.[3] As a compliment to the First Lord of the Admiralty, Sir James Graham, he named it 'Graham Island': there's flattery. A second Union Jack was planted when Alick Osborne, surgeon of HMS *Ganges*, and other officials from Malta landed on 20 August.

The birth of the volcano coincided with the arrival in Sicily of Ferdinand II, the new Bourbon King of the Two Sicilies. 'Never was the arrival of any monarch signalled by a more remarkable incident', Carlo Gemmellaro wrote, 'an event of which the whole Earth will register the history, and volumes of science will loudly proclaim an *event*, which I will dare to say Ferdinand himself might envy'. Gemmellaro named the island 'Ferdinandea'. So the island now had three names. The Spaniards raced towards the steaming island, too: they called it Nerita, a fourth name. The island had burst through the surface of the Mediterranean at a pivotal moment in European history: that year the Reform Bill was fought over on the streets of Britain and in Parliament, cholera swept much of northern Europe, and Michael Faraday discovered electromagnetic induction. For France, Prévost took full account of the significance of the island's birth exactly a year after the July Revolution had established Louis-Philippe as King of the French.

A new island, about halfway between Sicily and the coast of Tunisia, would be a rich strategic prize, potentially able to command the sea route between Gibraltar and Egypt without the inconvenience of the politically complex Malta. That is what the British, French, Sicilians and Spanish thought. National rivalry induced press hysteria, *The Times* reporting on a French newspaper's announcement of an English vessel being swallowed by a whirlpool off Graham Island.[4] A squib in the journal *John Bull* of September 1831 referred to the alleged appointment by the Prime Minister Earl Grey of his son as governor of Graham Island, and that he would take up his post as soon as the island had cooled down.

So it was to put the event into its historical, political and geological context that the distinguished mineralogist Carlo Gemmellaro addressed students at the Royal University of Catania on 28 August 1831, two months after the island had begun to make its presence felt. Gemmellaro told his students that the sea was speckled with floating pumice, and 'clouds of vapour charged with cinders were thrown out in the form of smoke, which became beautifully white in proportion as it rose.'[5] Some even more evocative and colourful notes came from the pen of John Davy:

> The bright light ultramarine hue of the sea underneath the dark brown Volcano canopied with sultry and snowy clouds of vapour of very great beauty and the surrounding sky where free from vapour of a similar hue less intense.[6]

Before the island sank back into the sea the following January, taking all conflicting national hopes and their flags with it, it became the most picked-over piece of real estate in the Mediterranean. The art dealer James Ackerman published a spirited lithographic sketch of the island by an unknown naval officer as early as September,[7] while another drew a set of precise watercolours of the island spitting and heaving as a party of sailors climb to the crater. Sir Walter Scott, by then an old and sick man, but generally regarded as a national treasure in both Scotland and England, landed on the island on 20 November when he was touring the Mediterranean in HMS *Barham*. He wrote about his brief exploration to the Royal Society of Edinburgh, of which he was president, and told how he sank up to his knees in the soft ash, and so climbed onto the shoulders of a stout sailor who carried him to the summit. His daughter Anne, who was accompanying him, burnt her shoes 'quite through' when picking her way across the hot sands.[8] On the island he saw two dolphins, 'killed apparently by the hot temperature', and a robin that appeared to have starved to death, and picked up a large block of lava and some shells for the Royal Society of Edinburgh.[9] But even Britain's own Prospero could not halt the geological processes working beneath the Mediterranean, and this magical island grew gradually smaller and smaller before disappearing altogether. Within a year it had become known as 'Graham Bank', and by 1841 it was 3 metres (ten feet) below the surface.[10]

The Napoleonic Wars were by now long over and, having demolished its rival navies, Britain ruled the seas. With a large navy, but nothing particularly warlike to do with it, the government looked for new uses for old sailors and their ships. Scientific exploration became one solution, and it was on a naval ship, HMS *Beagle*, that the young Charles Darwin set sail from Plymouth on his five-year voyage of natural history exploration in December 1831. Other expeditions transported and backed by the Admiralty included Ross, Parry and Franklin's searches for the North-West Passage between the Atlantic and the Pacific Oceans in the 1820s to 1840s, and James Clark Ross's expedition to the Antarctic, from 1839 to 1845.

An early result of Darwin's *Beagle* voyage was the publication in 1844 of his book *Volcanic Islands*.[11] 'I am quite charmed with Geology,' he wrote from the Falkland Islands in 1834, 'but, like the wise animal between two bundles of hay, I do not know which to like best; the old crystalline group of rocks, or the softer more fossiliferous beds.'[12] Covering the volcanic islands that Darwin visited, from the Cape Verde archipelago off West Africa, to Tahiti, the Galápagos and New Zealand, the book is one of a clutch of important works that put volcanoes at the centre of the scientific and literary stage. Two of these books brought geology into the modern world, while a third made a drama out of a crisis.

In the late eighteenth century, two theories of the origin of the earth were under vigorous debate: Neptunism and Plutonism. The former held the view that the earth is cold, and covered by sea, from which mountains were formed by precipitation and sedimentation. This view, propounded by the German mineralogist Abraham Werner (1750–1817), echoed the book of Genesis, and further suggested that volcanoes were merely burning coal deposits. Plutonism, on the other hand, put forward by James Hutton (1726–1797), urged that the earth had a molten core under high pressure, which volcanoes and earthquakes relieve. George Poulett Scrope's *Considerations on Volcanoes*, first published in 1825 when he was only 28 years old, favoured Plutonism. Scrope's book was the result of youthful research in the volcanic areas of France, Germany and Italy, during which Vesuvius made a timely eruption in 1822. Scrope put forward the theory that the earth was hot, cooling and alive with geological movements that produced earthquakes, tectonic plate shifting, volcanoes and huge waves. His own energies took him temporarily away from geology and into Parliament, where he was the silent member for Stroud for more than 30 years: 'a parliamentary reputation is like a woman's', he wrote: 'it must be exposed as little as possible'.[13] Scrope beat into print the eminent geologist Charles Lyell, whose *Principles of Geology* appeared in two volumes in 1830 and 1832. This proposed the theory of 'uniformism' which allowed that the earth had been cooling and forming over an immense length of time, creating volcanoes to relieve the pressures. Lyell was also quietly spoken, for all his skill as a lecturer: Emma Darwin, the wife of Charles Darwin, observed that 'Mr Lyell is enough to flatten a party, as he never speaks above his breath, so that everybody keeps lowering their tone to his.'[14]

These two quiet men set the tone for nineteenth-century studies of volcanoes. A third influence was the vain, effete, aristocratic and noisy figure of the novelist Edward Bulwer-Lytton. By the time he came to write *The Last Days of Pompeii* (published 1834) Bulwer Lytton had been described as 'without doubt, the most popular writer now living'.[15] Highly prolific and with an eye for a lavish storyline, Bulwer Lytton began his novel of highwaymen, *Paul Clifford* (1830), with the phrase 'It was a dark and stormy night', thus introducing a new cliché into English storytelling. Visiting Italy in 1833 with his wife, with whom he had a violent and volcanic marriage, Bulwer Lytton came home five months later with the manuscript of *The Last Days of Pompeii* in his bag. He and his wife promptly separated.[16] This dramatic tale of daily life in Pompeii, with its decadence, extremes of wealth and poverty, a story of forbidden love, and impending Christian martyrdom, reached a climax with the convenient eruption of Vesuvius and the destruction of the city. The story was not entirely Bulwer Lytton's own, however: the fate of Pompeii had been a melancholy theme in European literature and drama since excavations began in the mid-eighteenth century. Notably, Giovanni Pacini's opera *L'ultimo giorno di Pompei* was already in the Italian repertoire, having been first performed in Naples in 1825.[17]

Bulwer Lytton's novel was only one manifestation of an upsurge in artistic interest in the subject, which developed as a direct result of the activities of Vesuvius in the 1820s. For painting it climaxed in the monumental canvas, four and a half by six and a half metres (15 x 25 feet), by the Russian artist Karl Briullov (1799–1852). Painted from sketches made at Pompeii in 1828, Briullov's *Last Day of Pompeii* (note, like Pacini's title, the singular 'day'), was exhibited in Rome and Florence in 1833 where Bulwer Lytton saw it, and then in the Louvre, before being hung in the Imperial Academy of Arts in St Petersburg. It is now in the State Russian Museum, St Petersburg. The huge painting threw a number of distinguished literati off their guard: Sir Walter Scott, visiting Briullov in Rome in 1833, was perplexed and overcome by it, as was Gogol when it came to St Petersburg, and Pushkin who was inspired to write a poem, 'Vesuvius's Throat':

> Vesuvius' throat has opened – the smoke puffs thickly out –
> The flames are wild unfurled – like banners in a battle.

Karl Briullov, *Last Day of Pompeii*, 1833, oil on canvas.

> The earth is stirring – from the staggering pillars
> The idols tumble down! Beset with fear, people
> Throng, old and young alike, beneath the fiery dust
> And running in a rain of stones they flee the city.[18]

Riding on the wave of interest in the subject, Bulwer Lytton's book became an instant bestseller, and was soon translated into ten languages, followed by theatrical dramatizations and, in the twentieth century, by Hollywood films. But where Poulett Scrope and Lyell had carried out their own researches on volcanoes, Bulwer Lytton had turned to Pliny the Younger, whose letters to Tacitus describing the AD 79 eruption were already well known. As a man who wrote his novels at tremendous speed, Bulwer Lytton found Pliny to be a rich seam to mine for his invention.[19] Pliny's account, of which excerpts were quoted in chapter Three, is as precise and particular a description of an eruption as any we might now wish for. Bulwer Lytton, nearly 1,800 years later, writes an account that is heavily dependent on Pliny.

First, the eruption: 'a vast vapour shooting from the summit of Vesuvius, in the form of a gigantic pine-tree; the trunk, blackness, the branches, fire!'[20] Then comes the earthquake: '[They] felt the earth shake beneath their feet; the walls of the theatre trembled: and, beyond, in the distance, they heard the crash of falling roofs.' Then, the ash:

> an instant more, and the mountain-cloud seemed to roll towards them, dark and rapid, like a torrent; at the same time it cast forth from its bosom a shower of ashes mixed with vast fragments of burning stone! Over the crushing vines; over the desolate streets; over the amphitheatre itself; far and wide, with many a mighty splash in the agitated sea – fell that awful shower!

Then, the cloud:

> The cloud, which had scattered so deep a murkiness over the day, had now settled into a solid and impenetrable mass.

> It resembled less even the thickest gloom of a night in the open air than the close and blind darkness of some narrow room.

Then, the fire:

> But in proportion as the blackness gathered, did the lightnings around Vesuvius increase in their vivid and scorching glare. Nor was their horrible beauty confined to the usual hues of fire; no rainbow ever rivalled their varying and prodigal dyes. Now brightly blue as the most azure depth of a southern sky; now a livid and snake-like green, darting restlessly to and fro as the folds of an enormous serpent.

Then, the receding waters:

> The sea had retired from the shore and they that had fled to it had been so terrified by the agitation and preternatural shrinking of the element, [and] the gasping forms of the uncouth sea things which the waves had left upon the sand.[21]

Then finally, the aftermath:

> to the eyes and fancies of the affrighted wanderers, the unsubstantial vapours were as the bodily forms of gigantic foes – the agents of terror and of death.[22]

Despite the long gap in time between the two accounts, Bulwer Lytton follows Pliny carefully, as if there had been no scientific illumination in the preceding centuries. Not only does Bulwer Lytton give us Pliny's pine tree, but for good measure he includes what he describes as 'an enormous serpent' and 'gigantic foes'. The language of Bulwer Lytton emerges from the classical myth and allusion of Pliny's age in which he was well versed, rather than the language of science of his own. This was probably beyond him. Thus a literary perspective based on sensational fiction and

Frédéric-Henri Schopin, *Last Days of Pompeii*, *c.* 1850, oil on canvas.

imagery became the foundation of the popular perception of volcanoes in the nineteenth century, while Scrope's and Lyell's scientific responses needed more time to grow.

In the meantime artists had a field day with the subject. Following in the footsteps of Briullov, and catching the viral inspiration of the theme of the final days of Pompeii from both the Russian artist and from Bulwer Lytton, the French painter Frédéric-Henri Schopin (1804–1880) completed his *Last Days of Pompeii* around 1850. On a much smaller scale than Briullov, but sharing the staggering and the tumbling that Pushkin identified, we have here a metaphor for the collapse of a civilization, revealed in all its horror in a disorienting hellish light. And well may Schopin allude to the collapse of civilization, living as he did in France in the late 1840s. A new and bloody revolution was sweeping across Europe, removing kings, unsettling emperors, changing the very face of democracy and prompting the well-travelled historian Alexis de Tocqueville to say in the last session of the constitutional parliament in the French Chamber of Deputies in 1848: 'we are slumbering on a volcano'.

The American marine painter James Hamilton (1819–1878) travelled to London in 1854–5 and 1869, where he studied the work of Turner, Clarkson Stanfield and Samuel Prout. Returning home to Philadelphia in 1855 to paint subjects such as Hampstead Heath and the Welsh coast, he also produced the extraordinary *Last Days of Pompeii* (1864) in which Turnerian elements of vortices, massed classical buildings and flaming explosion are mixed with the gestural impressionistic manner that led to Hamilton being known as the 'Romantic Impressionist'.[23] The central column is more a memory of Nelson's Column in Trafalgar Square, which Hamilton would have remembered from his visits to London, than anything he might have imagined in Italy.[24] Hamilton is a transitional figure, highly competent and well-regarded, but falling between early nineteenth-century traditions, with their roots in British art, and the staunchly American Hudson River School of the middle and later years of the century. A perceptive reviewer suggested in 1847 that if Hamilton 'could continue to throw in a little more nature – American nature – and a greater air of reality . . . he might make magnificent pictures'.[25]

The year after Hamilton exhibited his *Last Days of Pompeii* in Philadelphia, back in London Edward Poynter RA (1836–1919) showed how universal was Bulwer Lytton's story of natural cataclysm, destruction and courage in his painting *Faithful unto Death* (1865). Its brave Roman subject, an example of true heroism such as was expected of British soldiers in the face of the enemy, was introduced in the Royal Academy exhibition's catalogue, thus:

> In carrying out the excavations near the Herculaneum gate of Pompeii, the skeleton of a soldier in full armour was discovered. Forgotten in the terror and confusion that reigned during the destruction of the city, the sentinel had received no order to quit his post, and while all sought their safety in flight, he remained faithful to his duty, notwithstanding the certain doom which awaited him.[26]

The literary source, however, is clearly Bulwer Lytton's description of a Roman sentry:

> the lightning flashed over his livid face and polished helmet, but his stern features were composed even in their awe! He remained erect and motionless at his post . . . he had not received the permission to desert his station and escape.[27]

This gap between artistic and literary image and scientific reality is one that becomes increasingly apparent during the eighteenth and nineteenth centuries. The size and nature of the gap varies according to scientific discipline, to the extent that it is perfectly clear that classical Greek sculptors' understanding of anatomy was far in advance of medical discovery, while in the nineteenth century artists lagged a long way behind scientists in the understanding of astronomy, for example. These are truisms, and can be easily explained, but it illuminates the fact that in art and science progress is a shared commodity, with obligations for cooperation thereby implied. As nature abhors a vacuum, so knowledge of the natural world grows by taking its nutrients from every direction.

It may reasonably be said that art shook science into getting ahead with volcanology. Who can look at images of a struggling Typhon trapped beneath Mount Etna, or contemporary engravings of the 1631 eruption of Vesuvius without experiencing a primitive fear of hideous death? By titling his book *The Vulcanos* in the seventeenth century, Anathasius Kircher's British editors put his subject at a measured distance, but then by giving it the subtitle *Burning and Fire-Vomiting Mountains* they underscored the visual reality of the subject. As every publisher knows, a good subtitle sells.

The difficulty we must face when looking at the eruptions painted by Wright, Volaire or Wutky in the late eighteenth century is that they take the viewer so close to the lava flow that the veracity of their paintings is immediately doubted. We want to believe it, but we cannot. The posing Rococo figures in Volaire's *Vesuvius Erupting at Night*, for example, painted *contre jour*, and

James Hamilton, *Last Days of Pompeii*, 1864, oil on canvas.

indeed *contre chaude rouge*, would or should be burnt painfully in moments. An English traveller, William R. Crompton, climbing Vesuvius in March 1818, remarked on the 'late eruption' that had taken place three months earlier in December 1817; but even so, by the following March,

> the surface is so hot and slippery you are obliged to hasten along to avoid being burnt . . . The suffocating power of the sulphur prevents you staying long . . . The different streams of lava are distinct, the colour ranging with time . . . and when cool become harder than most stone, the surface is black and resembles an agitated sea, each stream perhaps 40 yards broad and 15 or 20 feet deep.[28]

Given the varying levels of courage, bravado and imaginative invention between one artist and another, what artists actually saw when it was safe enough for them to climb Vesuvius was not always a livid pyroclastic flow, but more usually crumbling lumps of pumice and ash, and steaming black seas of cooling lava. In the late eighteenth century that would not have made a particularly appealing picture.

So it is instructive to see that as scientific knowledge of volcanology increases in the first two or three decades of the nineteenth century, so paintings of volcanic activity begin to become actually duller. When the Duke of Wellington was confronted by a clearly defined painting of the Battle of Waterloo by George Jones RA, he commented 'Not enough smoke'. He might well have said the same thing of a contemporary eruption painting, had he experienced volcanic as opposed to military violence. A transitional figure is the Norwegian artist Johan Christian Dahl (1788–1857), who stands between Wright of Derby or Volaire and Frederic Edwin Church (1826–1900), the American painter who travelled in South America in the 1850s. Dahl climbed Vesuvius in 1820 and painted a number of studies and at least two versions of a very smoky eruption. This takes an oblique view of the subject by obscuring much of the flank of the mountain behind smoke, and showing the lava collecting in a pool

J. C. Dahl, *Eruption of Vesuvius*, 1820, oil on canvas.

somewhere in the gap between the cone and Monte Somma, the volcanic 'collar' of Vesuvius. The foreground is filled with a wilderness of broken lava, clearly reminiscent of Crompton's 'black . . . agitated sea', while the intrepid figures remain at a tolerable distance from the heat. Dahl wrote in his diary that on 20 December 1820 he had climbed Vesuvius, 'and watched in daylight as well as in the evening, an important eruption – very interesting'.[29]

There is a hint of a new scientific approach here, rather than a literary and dramatic one, which is underlined by further notes made by Dahl in which he refers to a picture of 'the small and big crater taken from the ascent of Vesuvius, for Mr Monticelli, a professor of mineralogy in Naples.'[30] When Clarkson Stanfield (1793–1867) watched Vesuvius erupt on 1 January 1839, at a lucky moment during a tour of Europe, he recorded smoke dense enough to satisfy even the Duke of Wellington. This is Vesuvius as it was, rather than Vesuvius as it might be hoped or imagined. Unlike Wright or Turner, who went to Naples in the hope of seeing Vesuvius erupt, it happened for Stanfield by chance: Turner had left Naples a few weeks early, missing an eruption, while Stanfield arrived in the vicinity later than he had expected, and witnessed one. He and his friends climbed the mountain on 31 December when it must have been rumbling, and 'as there was

Clarkson Stanfield, *An Eruption of Mount Vesuvius*, 1820, watercolour and gouache.

a great deal of fire we had determined to stay till night to see the effects better, but the guides from certain symptoms thought proper to hurry the party down'.[31] Seven hours after they had got down the mountain it erupted spectacularly, but even after that narrow escape, Stanfield insisted on going back as far as the hermitage halfway up the mountain, where he saw the lava: 'the scene altogether was the most wonderful and sublime you can conceive. We remained as long as it was prudent.'

Across the first half of the nineteenth century, the high level of activity of the earth's volcanoes continued. Eruptions in the West Indies, the Far East, South and Central America and Iceland came with an unimaginable level of violence, some such as Tambora, with a power that matched both Santorini in 1620 BC and Vesuvius in AD 79 and 1631. Comfortably distant though they were, reports of these events gradually filtered through to Europe. While Tambora and the Icelandic volcanoes have been considered to be the greatest suppliers of atmospheric pollution

Eruption of Cotopaxi, 1741, engraving.

to Europe in the period, others in the western Pacific across a period of only twelve years, 1808 to 1822, contributed significantly to the burden of ash in the skies, and to the intensity of the colour of sunrises and sunsets: Agung (Bali): 1808; Mayon (Philippines): 1814; Tambora (Java): 1815; Raung (Java): 1817; Usu (Japan): 1822; and Galunggung (Java): 1822.[32] One serendipitous eruption, of Mount Erebus in Antarctica, was taking place in 1841 just as the first human visitors arrived on the continent, while others, such as Stromboli in the Mediterranean and Izalco off the coast of El Salvador, were in almost constant eruption. Izalco, born in 1770, came to be known as 'the lighthouse of the Pacific', until activity ceased in 1958. Such terrifying and seemingly continuous activity as that in the western Pacific in the first decades of the nineteenth century is a world away from the calm views of Mount Fuji and its landscape surroundings made by Utagawa Hiroshige (1797–1858) in the last few years before his death.

Nearly three decades before Darwin made his *Beagle* voyage, the German scientist Alexander von Humboldt (1769–1859) sailed to Central and South America, where between 1799 and 1804 he observed Andean volcanoes such as Cotopaxi and Sangay in both eruptive and quiescent states. Humboldt began to surmise a direct connection between earthquakes and volcanic activity, reaching the inaccurate conclusion that a network of interconnected geological events brought about the series of earthquakes that destroyed Caracas, Venezuela, in March 1812, those in the Mississippi basin in February 1812, and the eruption of Souffrière in St Vincent, West Indies, in April 1812. When Cotopaxi, in Ecuador, erupted in January 1803, Humboldt and his travelling companions were 290 km (180 miles) away at Guayaquil, where they heard 'day and night the noises proceeding from it, like the discharges of a battery [of cannon]'.[33] Humboldt's writings, in particular his *Cosmos* and *Personal Narratives of Travels to the Equinoctial Regions of America* (1845–62), inspired Frederic Edwin Church, the energetic and successful young American landscape painter, to travel to South America: 'Are we not justified,' Humboldt wrote,

> in hoping that landscape painting will flourish with a new and hitherto unknown brilliancy when artists of merit shall ... be enabled far in the interior of continents, in humid mountain valleys of the tropical world, to seize, with genuine freshness of a pure and youthful spirit, on the true image of the varied forms of nature?[34]

In 1853, and again in 1857, Church followed Humboldt's route in Colombia and Ecuador, where he made studies of the volcanoes Chimborazo, Cotopaxi and Sangay. Sketching Sangay

Frederic Edwin Church, *Cotopaxi*, 1862, oil on canvas.

from a camp at 4,000 metres (13,000 feet), he described an intermittent sequence of eruptions:

> At intervals of nearly four or five minutes an explosion took place; the first intimation was a fresh mass of smoke with sharply defined outlines rolling above the dark rocks followed by a heavy rumbling sound which reverberated among the mountains.[35]

From the foundation of these studies Church painted a series of monumental canvases including *Cotopaxi* (1862), *The Heart of the Andes* (1859), *Chimborazo* (1864) and *Rainy Season in the Tropics* (1866). All celebrated the sublime wildernesses of South America, a *terra nova* for American artists and collectors, who still looked east to Europe for the sources and wellsprings of their art, even as hundreds of thousands trekked west across the plains towards California.

Cotopaxi was commissioned by the collector James Lenox, the owner of the first two paintings by Turner to cross the Atlantic, *Fort Vimieux* (1831) and *Staffa, Fingal's Cave* (1832). While the lurid reds and oranges of *Cotopaxi*'s sunset are patently 'Turnerian' colours, and the composition and tenor of the work has been likened to Turner's *Fighting Temeraire* (1839), Church was making the painting during the first year of the American Civil War, and perhaps for this reason alone it can be read as a metaphor for the cruel, violent and sudden eruption of a peace-loving nation.[36] However, when it and *Chimborazo* came to London in 1865, Church was praised by the *Art-Journal* in terms that struck a decidedly fresh note:

> At length, here is the very painter Humboldt so longs for in his writings; the artist who, studying, not in our little hot-houses, but in Nature's great hot-house bounded by the tropics, with labour and large-thoughted particularity parallel to his own, should add a new and more magnificent kingdom of Nature to Art, and to our distincter knowledge.[37]

What caused Church to chime so directly with Humboldt was the German scientist's assertion in his *Cosmos* (first volume published 1845) of the intricacy of the connections throughout nature, and the interdependence of one part – geology, plant life, animals – upon another. Church wrote home in high excitement describing the flora and fauna, the cactuses taller than trees, multitudes of birds, the huge wasps' nests 'some as large as bushel baskets' and the overwhelmingly rich abundance of nature with which the landscape teemed.[38] Humboldt's evocation of the vastness of the earth, of its age, and of the eternal rhythms of nature that he experienced in South America, gave Church the intellectual foundation for his own 2.4-metre (8-foot) canvases. With Church's paintings of *Cotopaxi* and *Chimborazo* the subject of volcanoes in art has fully emerged from its literary stranglehold and rococo disdain of scientific reality. Pattern is no longer dominant over principle.

John Ruskin, fully at home when talking of art, but discomfited when addressing scientific matters, used volcanoes to create a metaphor for his own anxiety when writing of geological turmoil in the fourth volume of *Modern Painters* (1856):

> The experience we possess of volcanic agency is not yet large enough to enable us to set limits to its force; and as we see the rarity of subterraneous action generally proportioned to

> its violence, there may be appointed, in the natural order of things, convulsions to take place after certain epochs, on a scale which the human race has not yet lived long enough to witness. The soft silver cloud which writhes innocently on the crest of Vesuvius, rests there without intermission; but the fury which lays cities in sepulchres of lava bursts forth only after intervals of centuries; and the still fiercer indignation of the greater volcanoes, which makes half the globe quake with earthquake, and shrivels up whole kingdoms with flame, is recorded only in dim distances of history; so that it is not irrational to admit that there may yet be powers dormant, not destroyed, beneath the apparently calm surface of the earth, whose date of rest is the endurance of the human race, and whose date of action must be that of its doom.[39]

It is this sombre reflective mood that also underscores John Brett's *Mount Etna from the heights of Taormina* (1870), and Ruskin's own watercolour of Etna (1874), with its soft silver cloud writhing innocently. The scientific intention of Brett's painting

John Brett, *Etna from the Heights of Taormina*, 1870, oil on canvas.

John Ruskin, *Etna from Taormina*, 1874, watercolour and gouache.

is revealed in the fact that in 1870 he was in Sicily as the official draughtsman on a government-sponsored expedition to observe the solar eclipse of 22 December 1870. For the expedition he made a series of detailed drawings of the sun's corona, and wrote observations of the current eruptive activity that were published in *Nature* the following year:

> There has been a sad falling off in the appearance of Etna. The grand wreath of steam that used to roll out of the crater at such stately leisure . . . suddenly ceased about three days ago, and left nothing more than a tiny wisp of smoke rather suggestive of a cottage chimney than a volcano . . . I have watched the volcano at all hours of the day for a week past, in the hopes of getting a correct outline of it for pictorial purposes. The clouds only cleared off completely yesterday.[40]

Brett shows Etna as the most distant landscape element in a painting that moves from cultivated orchards, to the peaceful town of Taormina, across low folded mountains and a fertile river valley, to the cold, inhospitable summit of the volcano. Thus Etna is placed in context, with only its soft silver cloud being allowed to suggest its potential for mayhem.[41] It is interesting that the 'cottage chimney' analogy that Brett used in describing Etna was of the same domestic tone as that drawn by William Hamilton over a hundred years earlier when likening Vesuvius in full spate to 'quivering like timbers in a water-mill' and flowing 'like the Severn, at the passage near Bristol'.

The progress of understanding takes place with jerks, reverses and reconsiderations. Rather than paint Pompeii in the full flow of its terminal disaster, the Italian realist Filippo Palizzi (1818–1899), a generation younger than Briullov and Schopin, shows it instead being dug out again. Later nineteenth-century artists such as Palizzi and Poynter garnered the social and architectural details in their paintings from archaeological digs in process in Pompeii, Egypt, the Middle East and elsewhere. Taking an alternative view of the subject in his *Study of an Excavation (Pompeii)* (1864) Palizzi gives us a midday calm, as the men with shovels have downed their tools and left the site temporarily to wildflowers and silence. Palizzi's compatriot, Gioacchino Toma (1836–1891) takes a different kind of modern view in his *Sotto al Vesuvio di Mattina* (1882). Here Vesuvius smokes gently, while a steam engine puffs energetically below it, travelling as fast as it possibly can. What vanity, suggests Toma, a man who had fought with Garibaldi to unify Italy and had been imprisoned for his pains; just wait until Vesuvius blows and then see how clever the steam engine is.

The understanding of volcanoes was not helped by authors such as Bulwer Lytton or indeed Jules Verne whose *Journey to the Centre of the Earth* (1864) takes us back to the view proposed by Athanasius Kircher exactly 200 years earlier, and to the experiences of Baron Munchausen in the late eighteenth century. Verne suggested that the volcano that launched his Professor Lindenbrook and nephew back to the surface of the earth was part of an interconnected warren of drains and channels running between the centre and the surface, which could allow the good

View of the square at Torre del Greco, Campania, during the eruption of Vesuvius on 8 December 1861, 1877, engraving.

professor to enter the system at Snaefells in Iceland, and eject from it in Stromboli, on the Lipari Islands.[42] Would that it could; it would be so convenient for intercontinental travel, though it would challenge the airlines, and demand an urgent solution to the problem of excess heat.

Despite the generational change of step clearly apparent 50 years after Dahl climbed Vesuvius and addressed the volcano as a scientific subject, it is salutary to note that the Queen's Theatre in Long Acre, London, was so out of touch in 1877 as to mount a dramatized version of Bulwer Lytton's *Last Days of Pompeii*. It was a resounding flop.[43]

6 Krakatoa Shakes the World

On 27 August 1883, the island of Krakatoa (Krakatau) in the Sunda Strait, midway between Java and Sumatra, exploded with a noise that woke the planet.[1]

The Krakatoa 'event', as volcanologists disarmingly put it, profoundly disrupting and disturbing though it was for the immediate area, was just one of many cataclysmic eruptions within historic memory that were of the scale Volcanic Explosivity Index (VEI) of 6.5. This was larger than Vesuvius in AD 79 (VEI 5.8), but smaller than Tambora in 1815 (VEI 7.3). It was still smaller than some prehistoric eruptions, and yet larger ones had taken place in the millions of years when the earth was still forming, but after animal life began. This is a well-shaken planet.

The volcanic Krakatoa island, which before 1883 had been known as the 'island with a pointed mountain', straddled the area of sea where the Indo-Australian tectonic plate, colliding into the Eurasian plate, is diverted downwards towards the earth's core. As a consequence, there is a line of active volcanoes along its leading edge. To say that the island was destroyed by the volcano would be misleading, as it and the sea basin around it had also been formed as a result of eruptions, and this was just one step in a continuous process. Volcanoes explode at points where pressure from simple mechanical shifts beneath the earth's crust is too strong for the crust to hold in place.

Until mid-summer 1883, the Sunda Strait was the highway for vessels trading east and west from Sumatra and Java. There had been preliminary events beginning around 10 May, when

Parker & Coward, *The Eruption of Krakatoa, 27 May 1883*, 1888, lithograph.

Eruption of the Volcano Perbouwatan, on the Island of Krakatoa, 1883, engraving.

rumblings were heard and a lighthouse was seen to move. Ships' captains filed anxious reports. Ten days later a thick white column of smoke and ash rose from the volcano to a height of about 11,000 metres.[2] *The Times* reported these events on 3 July, but they turned out merely to be polite preliminaries.

The sound of the main cataclysm of late August radiated many thousands of miles around the Pacific, being heard on the west coast of Australia to the southeast, and Rodriguez Island and Diego Garcia to the west. A tsunami followed this huge breaking wave of sound, falling eventually on Ceylon and the east coast of India, and making tide-gauges flicker as far away as Port Elizabeth in South Africa, Grytviken in South Georgia and Biarritz in France. Even in Birmingham, England, the calm rustle of the trees at Edgbaston were disturbed as the local observatory registered changes in air pressure as the shock-waves ran around the globe.[3] Noise and tides pass very quickly, and leave no trace. What did remain, however, and for months, was a cloud of ash dust and gas that encircled the planet in the upper atmosphere: 'where mountain Krakatua once stood, the sea now

SEPTEMBER 29, 1883. HARPER'S WEEKLY. 6

THE ISLAND AND VOLCANO OF KRAKATOA, STRAIT OF SUNDA, SUBMERGED DURING THE LATE ERUPTION.—[SEE PAGE 614.]

the University, and of other talented persons, in whose presence the astonished nobility of Suabia raised its chin and sniffed.

All these thronged the sunny drawing-room on the second floor of the "Englische Palais"; and the reflection that there was safety in numbers did not prevent me from being silently jealous of each and every one of them. My own position was neither dignified nor agreeable. Being in some sort a member of the household, I was used by Lady Constance much as the wives of certain generals are in the habit of using their husbands' aides-de-camp. I was ordered to ring the bell; I was sent to look for this or that person's carriage: I fetched and carried, in short, and got nothing in the shape of reward, except a smile or a word every now and again. Sometimes I thought that she took a delight in humiliating me, though I can not say that her manner betrayed any such deliberate intention. I did not like it; but I grinned and bore it, sustained by an inward conviction that, sooner or later, my would come.

A circumstance which vexed me a good deal was the sud appearance upon the scene of Mr. Sotheran, who took up his qu ters in the best hotel in the town, and immediately resumed th patient siege operations which had for their object the ultim conquest of Lady Constance's hand and heart. Little as I lo this dogged wooer, I could not refuse a grudging tribute of miration to his unruffled pertinacity. It was not often that he t umbrage as he had done at Taormina; but then it was not of that Lady Constance laughed at him. She turned him away fr her door full twice for every once that she admitted him; snubbed or took no notice of him when he had succeeded in eff ing an entrance; but he cared for none of these things. He me to win, and really it almost looked as if he would win in the lo

'The Island of Krakatoa, Strait of Sunda, Submerged during the Late Eruption', from *Harper's Weekly* (1883).

plays', wrote *The Times*.[4] The dust was the pulverised mountain floating silently around the earth, creating lurid and intense sunsets and sunrises. The particles scattered the sun's light, dispersing the shorter blue/violet end of the spectrum and allowing the longer red wavelengths to dominate. These beautiful effects of volcano-produced light had not been experienced within living memory, as the sunset effects of the post-Tambora months in 1816 were no longer within that span, except among observant septuagenarians blessed by life and chance. None, however, seems to have mentioned it.

The post-Krakatoa effects may well have been magnificent in London, but as the London atmosphere was already heavily polluted by industrial and domestic smoke, the Krakatoa eruption cannot have had such a marked effect. London fogs had been notorious at home and abroad for their filth, texture, smell and danger since the early decades of the century. A French visitor called London 'Brouillardopolis',[5] and this was not because of post-Krakatoa effects. Nor was it the reason that Claude Monet came to paint the London Thames in 1899, 1903 and 1904: the pictorial attraction for Monet was the sun seen through the fog, where

> [Waterloo] bridge slumbers in ever deepening tones of blue with long melting greenish reflections. A sulphurous crimson ray drags in the colour of the River Styx and smoulders beneath an arch where its glint turns to carbon ... A boat glides or rises up like a violet shadow.[6]

The centre and East End of London was already a many-hued palette for painters years before and after Krakatoa. A few miles further west, in Chelsea, however, where the air was clearer, the sky registered the special effects of the Krakatoa eruption as expressed by the local painter William Ascroft.

Ascroft was a prolific watercolour painter specializing in views of the Thames. He may have been aware as a growing boy of the presence in Chelsea in the 1840s of a mysterious old man who pretended to be one Captain Booth, a seaman, and spent considerable time being rowed out onto the river. This was in fact the painter J.M.W. Turner, who lived in Davis Place with his companion the widow Mrs Sophia Booth until his death in 1851. In the 1860s and '70s James McNeill Whistler lived in Chelsea, where it is likely that Ascroft knew of him. Atmospherics were well studied by artists in Chelsea, and so it may not have been so surprising for one to be seen drawing the evening skies around Chelsea in pastel, day in, day out, in the autumn of 1883.

Ascroft does not seem to have been aware of a distinct beginning to the eruption sequence, but may have been drawing the Chelsea skies as a matter of course and, in retrospect, realised that something unusual was going on. When he looked back at his sketches later, Ascroft realised that two of them, of 8 September, showed an unusual afterglow, which by 20 September, a month after the eruption, was 'indisputable'.[7] The first strong afterglow he observed and recorded on 8 November, 'when a lurid light was seen about half an hour after sunset. It was so extraordinary that some fire engines turned out'.[8] There were yet more extraordinary manifestations: the sun appeared to have no beams, which at sunset on an average cloudy day can be magnificent. Further, as Ascroft put it,

> the light was so sickly that clouds opposite the sun had none of that richness of colour which is usually seen at, or just after, sunset. In the daytime the lights on clouds under the sun were frequently of a greenish-white hue.

William Ascroft, three sketches from *Twilight and Chromatic Afterglow, Chelsea*, November 1883–6, pastels, reproduced lithographically, 1888.

So it went on, and until April 1886, Ascroft continued to record the phenomenon. He noticed what came to be called 'Bishop's Rings', distinct bluish or brown circles around the sun caused by volcanic dust in the atmosphere, and what Ascroft described as 'blood afterglows' and 'amber afterglows'. By July 1888, when the drawings were exhibited as a group of more than 530 at the South Kensington Museum (the original name of the Victoria and Albert Museum), the artist recorded that the afterglows 'are still occasionally seen, though much modified'.[9] This was five years after the eruption. We should observe a note of caution, because although with 'few exceptions the sketches were done directly from Nature', Ascroft admits that some of the drawings were done from memory. They were, however, considered to be of such interest that selections were exhibited at conversazioni of the Royal Society.[10] When the drawings came to be printed for the Krakatoa Committee of the Royal Society Ascroft insisted that the printers worked only in bright weather to avoid the errors that he considered had been made in printing colour reproductions of Turner's paintings.[11]

Some 320 km (200 miles) northwest of Chelsea, at Stonyhurst College, Lancashire, another especially literate observer, the poet Gerard Manley Hopkins, looked carefully at the autumn and winter skies in 1883. He collected evidence from scientists around the world, and also wrote to the editor of the journal *Nature* about what he himself had seen:

> The glow is intense, this is what strikes everyone; it has prolonged the daylight, and optically changed the season;

Twilight and Afterglow Effects at Chelsea, London.
Nov. 26th 1883.
No. 4. About 4.40 p.m.
No. 5. About 5 p.m.

> it bathes the whole sky, it is mistaken for the reflection of a great fire; at the sundown itself and southwards from that on December 4, I took note of it as more like inflamed flesh than the lucid reds of ordinary sunsets. On the same evening the fields facing west glowed as if overlaid with yellow wax.
>
> But it is also lustreless. A bright sunset lines the clouds so that their brims look like gold, brass, bronze or steel. It fetches out those dazzling flecks and spangles which people call fish-scales. It gives to a mackerel or dappled cloudrack the appearance of quilted crimson silk, or a ploughed field glazed with crimson ice. These effects may have been seen in the late sunsets, but they are not the specific after-glow; that is, without gloss or lustre.
>
> The two things together, that is intensity of light and want of lustre, give to objects on the earth the peculiar illumination which may be seen in studios and other well-lit rooms, and which itself affects the practice of painters and may be seen in their works, notably Rembrandt's, disguising or feebly showing the outlines and distinctions of things, but fetching out white surfaces and coloured stuffs with a rich and inward and seemingly self-luminous glow.[12]

Reds, yellow, gold, brass, bronze, steel, crimson: these are the tones and colours that Hopkins saw in the sky. And 'glow', 'reflection', 'inflamed', 'dazzling flecks and spangles' and 'self-luminous glow': these are among his epithets. No wonder he invoked Rembrandt in his analysis of the northern English skies, because these were, after Krakatoa, extraordinary, and never so lucidly recalled.

In his novel of 1889, *Blown to Bits*, R. M. Ballantyne wrote: 'It is no figure of speech to say that the *world* heard that crash. Hundreds, ay, thousands of miles did the sound of the mighty upheaval pass over land and sea to startle, more or less, the nations of the earth.'[13] Tennyson, in his poem 'St Telemachus' (1892), used the spectacular reports of the skies to open his poem about the martyrdom of Telemachus in the Roman Colosseum:

Had the fierce ashes of some fiery peak
Been hurl'd so high they ranged about the globe?
For day by day, thro' many a blood-red eve . . .
The wrathful sunset glared.

As Bulwer Lytton had used the eruption of Vesuvius as the background to *The Last Days of Pompeii*, so Tennyson pulled the eruption of Krakatoa back in time to create a suitably apocalyptic context for the death of the anchorite Telemachus, 'bathed in that lurid crimson'. Krakatoa was not Tennyson's first volcano background. He had described Kilauea in Hawaii in his poem about the Christian woman Kapiolani defying the pagan god of the volcano, Pelé:

Long as the lava-light
Glares from the lava-lake
Dazing the starlight,
Long as the silvery vapour in daylight
Over the mountain
Floats, with the glory of Kapiolani be mingled with either
  on Hawa-i-ee.

The *New York Times* reported that the colours in the sky above the eastern seaboard of the United States were so bright in red and purple, that some were convinced that they saw the Stars and Stripes carpeting the heavens:

> Soon after 5 o'clock the western horizon suddenly flamed into a brilliant scarlet, which crimsoned sky and clouds. People in the streets were startled at the unwonted sight and gathered in little groups on all the corners to gaze into the West. Many thought that a great fire was in progress . . . The sky that morning was fairly aglow with crimson and golden fires, when suddenly, to their great astonishment, an immense American flag, composed of the national colours, stood out in bold relief high in the heavens, continuing in view for a considerable length of time.[14]

Edvard Munch, *The Scream*, 1893, oil, tempera and pastel on cardboard.

Frederic Church, at Olana, his Moorish-style hilltop mansion overlooking the Hudson Valley, soon heard about the drama of the skies and made a measured response. He wanted to see the fiery skies in a bleak, deserted environment, and travelled north-west to Chaumont Bay on the shore of Lake Ontario. There, in late December, he painted the pack ice on the lake below the lurid sky of pink and purple, orange and mauve.

In Norway, Edvard Munch wrote of 'Clouds like blood and tongues of fire hung above the blue-black fjord and the city.'[15] Ten years later he came to paint *The Scream* (1893) in which not only is the man on the bridge screaming, but the whole sky and the landscape below it.

The Royal Society brought together a committee to assemble information about the Krakatoa eruption, publishing in *The Times* and elsewhere in February 1884 a request for accounts of ash fall, pumice, barometric phenomena and 'exceptional effects of light and colour in the atmosphere' to be sent to them.[16] The committee reported in 1888. Ascroft's awareness on 8 September 1883 that something odd was happening was later corroborated exactly by the Hon. Rollo Russell, a member of the committee living in Surrey. He looked back at his diary on 9 November when he first noticed that the skies were displaying 'a very striking or extraordinary character'. He saw that on 8 September he had observed 'a fine red sunset with after-glow', adding: 'this is worth remarking, because I had never previously used the expression "after-glow"'.[17] Thereafter, he noted a kaleidoscope of colourful effects at sunset: on 26 September 'light pink cirrus stripe'; 3 October 'red and yellow sunset'; 20 October 'fine reddish sunset with bright isolated cloud and slight low cirrus'. The intensity of effect began to increase its pace the following month when Russell further noticed bright green opalescent light, green haze, and bright opalescent yellows. In early March 1884 he reported 'a slight repetition of the sky-illumination, lasting only 30 minutes, but during March the glare completely vanished, and no illumination whatever appeared in a clear sky after sunset. During the remainder of the year the sunsets were uncommonly free from colour.'

Reports came to the committee from San Remo, Cannes, Berlin, Lisbon, Japan, New Caledonia, San Salvador, Panama, Transvaal and the USA, and as a gloss on his rich and fulsome report from Surrey, Rollo Russell compiled a table of 'Previous analogous glow phenomena, and corresponding eruptions', gathering together reports of atmospheric phenomena from as far back as the eruption of Pichincha in 1553, which, according to the seventeenth-century Danish historian Jens Birkerod, was the cause of 'remarkable purple after-glow in Denmark, Sweden and Norway'.[18] Thus faded the most extraordinary sequence of sky-effects seen across Britain, in which the colours of Fauvism and Expressionism danced across the northern skies.

To those on the spot, the eruption of Krakatoa did not just mean extraordinary evening skies, but an experience of hell on earth, as one terse eyewitness, the captain of the British ship ss *Charles Bell*, wrote in his log:

> At 5 [22 August 1883] the roaring noise continued and increased, wind moderate from SSW, darkness spread over the sky and a hail of pumice stone fell on us, many pieces being of considerable size, and quite warm ... The continual explosive roars of Krakatoa, made our situation a truly awful one ... The sky one second intense blackness and the next a blaze of fire, mastheads and yardarms studded with corposants [phosphorescent electrical discharges, St Elmo's Fire], and a peculiar pinky flame coming from clouds which seemed to touch the mastheads and yardarms ... The ship from truck to waterline is as if cemented, spars, sails, blocks, and ropes in a terrible mess but thank God nobody hurt or ship damaged. On the other hand how fares it with Auger, Merak, and other little villages on the Java coast.'[19]

The year 1883 also marked a break that was quite as shattering for European culture as it was for Pacific geology. A decade earlier, Impressionism had arrived as a flash of discontent with salon art in Paris. In 1883 the nineteenth-century musical revolutionary Richard Wagner died, and these crucially twentieth-century figures were born: the architect Walter Gropius, the writer Franz Kafka, the demagogue Benito Mussolini, and the leader of fashion Coco Chanel. Wagner's death marked the end of one cultural period; the births of Gropius, Kafka, Mussolini and Chanel the many beginnings of quite another. While artists' depiction of landscape would alter radically after Impressionism, so after Krakatoa the depiction of volcanoes would change. This was not as a result of the eruption itself, but one more symptom of violent social, artistic and technological changes that found their fault lines beneath Paris, London and New York in the late nineteenth century. Krakatoa, beneath whose ashes and tsunami 36,000 people died, thus becomes an evocative symbol.

Underwood & Underwood, Enormous smoke column, three miles high, mushrooming over Mount Pelée during the eruption, June 1902, photograph.

7 'The Night had Vanished': Vorticism and the Volcano

The first volcanic eruption of scale and substance in the twentieth century shook the West Indies in May 1902. The Windward islands of Dominica, Martinique, St Lucia and St Vincent, which run due south in a thin line towards the coast of Venezuela, burned and broiled as their volcanoes went off like firecrackers on the tail of a cat. They and most of the West Indies chain are a system of volcanoes that runs along the line of the relatively small Caribbean Plate as it meets the South American Plate. The town of Saint-Pierre on Martinique was completely destroyed by the volcano Mount Pelée, with comparatively lesser damage done on St Lucia and St Vincent when their volcanoes, both named Souffrière, erupted in sympathy. At home in England, responding to press reports, the diarist Mary Monkswell wrote these eloquent words on the terrible events:

> A second Pompeii. On this awful day the island of Martinique, a French colony in the W. Indies was, after a long threatening but literally at about 5 minutes notice, completely enveloped by burning lava and mud from the volcano La Pelée, and some thirty to forty thousand people at once suffocated. All the ships in the harbour were at once set fire to and sunk except one, the *Rhoddam*, an English steamer, whose steam was fortunately up. The Captain (Freeman), a Glasgow man, whose hand I should like to shake, in spite of seeing half his crew fall dead and his ship on fire in 40 places, and frightful burns and injuries to himself, had the

anchor chain cut and steered her through thick darkness out of that awful bay.[1]

Despite Mary Monkswell's understanding that there had been no notice, there had been many warnings since January about this sudden eruption in the elegant Saint-Pierre, known locally as the Paris of the West Indies. In late April, earthquakes on the slopes of the mountain and the rising heat and gases on its flanks drove hordes of insects, snakes and animals into Saint-Pierre where they terrified the population and led to 50 or more deaths. Mindful of a coming election the governor used the local press to soothe the anxious population back to the city, and departing refugees were sent back to Saint-Pierre by road blocks manned by troops. But then, of course, came the eruption. It was smaller than Krakatoa, but that was of no comfort to the 28,000 inhabitants. There were only three survivors: one was a prisoner, Louis-Auguste Cyparis, who had been locked up in solitary confinement for grievous bodily harm, and was protected by the security of his cell. He spent his remaining days as a performer in Barnum & Bailey's Circus, celebrated as 'the Man who Survived Doomsday'.

The eruption brought a crop of shuddering accounts not only from people who happened to survive on the few ships that escaped from the harbour, but also by those who had witnessed other eruptions, such as the Revd Charles G. Williamson, who had been present in 1868 when Mauna Loa in Hawaii erupted. 'At the present moment a few details of that appalling circumstance may be of general interest', he wrote to the editor of *The Times*.[2] What good can possibly come out of a tragedy such as the 1902 eruption of Mount Pelée, except the tales of a garrulous circus freak, the reminiscences of a church missionary and the promise of a fertile landscape a generation or two later?

Comprehensive, sudden devastation with the destruction of life and collective memory brings an emptiness that, in time, can only be evoked by art. Patrick Leigh Fermor's novel *The Violins of Saint-Jacques*, published in 1953, 50 years after the Saint-Pierre tragedy, found poetry in the disaster. He told the story of the

Cover design for the first edition of Patrick Leigh Fermor, *The Violins of Saint-Jacques* (1953).

mythical island of Saint-Jacques, situated, according to Leigh Fermor, between Guadeloupe and Dominica, in the Windward Islands, 'where it hung like a bead on the sixty-first meridian'.[3]

Set in the last few years of the nineteenth century, the story is seen through the eyes of a young painter Berthe de Rennes, who describes the delights and luxuries, customs, courtesies and venalities of the island in the days before a ball at Government House. Berthe drew and painted on the island, and sent albums of her pictures back to her aunt in Paris. As hams and quails in aspic, giant lobsters and crabs, ivory pyramids of *chou coco* and mounds of soursop and mangoes were assembled for the festivities, 'the volcano [the Saltpêtrière] had been burning . . . with unaccustomed vigour. Now it hung in the dark like a bright red torch, prompting the island wiseacres . . . to shake their heads.'[4] At the height of the ball, when dancers were whirling and the orchestra reaching a frenzy, panic set in as a group of lepers disguised in pantomime domino costumes, intermingled with the revellers. This fear was soon overwhelmed by a far greater terror as the volcano erupted with unparalleled violence during the Governor's firework display:

> The night had vanished. Everything was suddenly brighter than noonday and from the crater of the Saltpêtrière a broad pillar of red and white flame, thickly streaked with black, was shooting into the sky like the fire from a cannon's mouth. It climbed higher every second until it had reached a fierce zenith miles up in the air, and the roar that accompanied its journey was interrupted by hoarse thunderclaps that almost broke the eardrum.[5]

Unlike Saint-Pierre on Martinique, but like uninhabited Krakatoa, Sabrina and Graham Island, the island of Saint-Jacques, 'its mountain and its forests, its beautiful town, and the forty-two thousand souls who had lived there'[6] all disappeared beneath the sea. By a flick of the novelist's pen, Berthe survived – as did her pictures, safely by now in Paris. Some time later fishermen would claim that anyone 'crossing the eastern channel between the islands in carnival time, can hear the sound of violins coming up through the water. As though a ball were in full swing at the bottom of the sea'.[7] The painting that roused the narrator's interest at the beginning of the story – 'the last thing I painted in the Antilles', said Berthe – was signed and dated 'B. de Rennes, 1902',[8] as significant a date for the fictional Saint-Jacques, as for the all-too-real Saint-Pierre.

The event in Martinique was a prelude to another major eruption of Vesuvius. Here artists were present, on or near the spot, and in some quantity. Where Saint-Pierre was evoked only by a novelist and his fictional artist, those present in or near Naples in April 1906 when Vesuvius erupted included the Italian painter Eduardo Dalbono (1841–1915) and an American artist resident in Capri, Charles Caryl Coleman (1840–1928). Coleman, born in Buffalo, New York, had travelled in Italy as a young man, studying in Florence and Rome, where from the late 1860s he settled in the area around the Spanish Steps, long popular with

European and American expatriate artists. He moved to Capri in 1885, where he made his home in the Villa Narcissus, a former convent gatehouse in which he recreated the contrasting atmospheres of Pompeii and Morocco.[9] Coleman became a leading figure among the artists on the island, sending regular consignments of pictures back to the United States, and in particular to the Brooklyn Museum, for exhibition and sale. Like Frederic Church, who had also created a house in the Moorish style filled with artefacts and languid artistic atmosphere, Coleman took full advantage of the views from his studio windows: Church had the Hudson Valley in front of him, while Coleman surveyed the Bay of Naples with Vesuvius on the horizon.

From his easel, and at approximately the same distance from the volcano that Pliny the Younger had been more than 1,800 years earlier, Coleman made a series of pastel drawings that record the slow progress of the eruption, and the billowing smoke. At this distance, the eruption is a silent, far-away decorative device, softly evoked in shades of blue and grey. The subtitle of one of Coleman's drawings, *A Shower of Ashes upon Ottaviano*, is a detached acknowledgement of the devastation of the small town, which became known as 'the new Pompeii'. By contrast, the ink and gouache study by Eduardo Dalbono, which must have been drawn on or near the same day as Coleman made his distanced view, takes the viewer right into the heart of the disaster, sharing the terror and the panic of the inhabitants as they flee with their belongings. While a religious procession makes its way through the town, a column of soldiers is marching towards the mountain. The artist's inscription evokes the horror and, in particular, the appalling noise: 'Ash, darkness, roaring, panic envelopes the road from Naples to Resina'[10] as workmen hopelessly shovel ash into baskets and others huddle pathetically under umbrellas. Far apart though they are in distance and mood, the Coleman and the Dalbono have the same title, *A Shower of Ashes*.

In the period when American artists were travelling thousands of miles into Europe and South America and evoking the landscape and manners they discovered there, Icelandic painters

Charles C. Coleman, *A Shower of Ashes upon Ottaviano*, 1906, pastel.

Eduardo Dalbono, *The Shower of Ashes – The Eruption of Vesuvius, 11 April 1906*, 1906, ink and gouache.

began slowly and with considerable logistical effort to travel mere hundreds to discover the interior of their own island. Icelandic artists born in the late nineteenth and early twentieth centuries travelled overwhelmingly to Copenhagen for their art training. Iceland until 1874 was governed from Denmark, and had no formal public art school until 1947 when the Reykjavík School of Visual Art was founded.[11] Ásgrímur Jónsson (1876–1958) studied at the Royal Academy of Fine Arts in Copenhagen in 1900–1903, as did Brynjólfur Thórdarson (1896–1938), Finnur

Jónsson (1892–1989), Gudmundur Einarsson (1895–1963) and Johannes Kjarval (1885–1972). While in Denmark, and on further travels into Germany, France and Italy, these young artists gradually came across Impressionism and Post-Impressionism, and following directly on from them they experienced the tornado of brewing European artistic styles, including Expressionism and Vorticism.

A small, rocky, cold and – compared to the rest of Scandinavia and Europe – formerly backward island, Iceland moved more slowly and cautiously than most nations into the twentieth century. But what Iceland had that was conspicuously absent in Denmark and every other European state except Italy was an active, menacing and to artists highly attractive system of volcanoes, with natural and plentiful heat and light. One of the largest and liveliest, Grimsvotn in the central highlands of Iceland, erupted in 1902–3, while Hekla, a wide-shouldered, low-domed oppressive mountain dominating the southwestern region, erupted in 1878 and again in 1913. Such volcanic activity was thus integral to the cultural heritage of the younger Icelandic artists of the twentieth century.

Watercolours by Ásgrímur Jónsson, painted soon after 1909 when he had returned from a twelve-year absence in Copenhagen, Germany and Italy, depict an eruption taking place in a fiery effusion of reds starkly contrasted against blues. While Turner's watercolours may be a source for some of these works, there is also, in the soaking of the paint into the paper, an evocation of Emil Nolde, and abstract shapes that recall Munch. But throughout, Ásgrímur maintains a strong narrative in the presence of the figures of country people looking on at the eruptions either aghast or in resignation. Whether they run from the explosions, look on them with fear or regard them with noble detachment, Ásgrìmur's figures have a staunch humanity that derives from the tradition of the Icelandic sagas.

Johannes Kjarval, some twenty years younger than Ásgrímur, was like him the son of a farmer in southern Iceland. Spending his late teenage years working on fishing boats, Kjarval came to know Ásgrímur, and to learn painting techniques from him and

Ásgrímur Jónsson, *Flight from a Volcanic Eruption*, 1945, oil on canvas.

Ásgrímur Jónsson, *Eruption*, *c.* 1908, watercolour.

Johannes S. Kjarval, *Mount Skjadbreidor*, view from Grafningur, *c.* 1957–61, oil on canvas.

another pioneering Icelandic painter, Thorarinn Thorlaksson. Kjarval travelled as a young man first to London in 1911, where he discovered Turner's paintings at the National Gallery and in the National Gallery of British Art, now Tate Britain. He moved to Copenhagen to study at the Royal Academy of Art, returning eventually to Iceland in 1922 to settle in Reykjavík. Kjarval's approach to the Icelandic landscape was to make large, intensely detailed canvases, in jewel-like colour, fragmentary forms and often very little sense of a horizon. The eye is therefore pulled straight up into the image with little external reference. The paintings have a harmony that reflects the artist's practice of painting out in the landscape, often while a taxi waited patiently for him a respectable distance away. Kjarval's volcanic subject-matter was not the violently exploding beast in the distance, but the cooled and cold rocky buckling lava that covers much of Iceland. 'All nature is a single symphony, all of it music', he wrote in 1954. 'You are so receptive to music out in the lava field.'

Gudmundur Einarsson studied at Stefan Eiriksson's school of art in Reykjavík, and with Ríkardur Jónsson and Thorarinn Thorlaksson, from 1911 to 1913. He moved on to study in Copenhagen and Munich. His *Eruption of Grimsvotn* (1934) is as viscerally shocking as Kjarval's reflective landscapes of eruptive aftermath are calm and regenerative. Using forms derived from Vorticism, and indeed also from the nineteenth-century German Romantic painter Caspar David Friedrich, Einarsson conveys with sharp diagonals and searing colour the sudden, destructive and overwhelming scream of the mountain as it explodes into smoke and fragments. Einarsson had experienced Icelandic volcanoes at first hand, being a pioneer of mountaineering, and the director of a film on the 1946 eruption of Hekla.[12] The chaos he evokes in his Grimsvotn painting is tightly controlled as he releases the fury of the earth in fire and billowing smoke. The softness of the smoke contrasts with the sharpness of the flame, and uncannily prefigures the mushroom cloud of an atomic explosion, eleven years before the world first saw such a thing. But the world did have the opportunity of seeing this painting some time before Icelandic art became more widely known, as it was exhibited at the Minnesota State Fair in 1940.

Twentieth-century Icelandic painters were just as important as pioneers in the subject of volcanoes as were European artists such as Volaire and Wright in the late eighteenth century. The latter encountered the erupting Vesuvius from the historical background of the picturesque and the Rococo, styles of charm and wit but of little emotional charge. Suddenly grasping the volcano's mysterious power, they produced images that defined their subject for a century, and indeed continue to do so. By contrast, Ásgrímur Jónsson, Gudmundur Einarsson, Johannes Kjarval

Gudmundur Einarsson, *Eruption of Grimsvotn*, 1934, oil on canvas.

and others came across these shattering natural phenomena in the light of their studies in Copenhagen, Germany and London, before, during and after the First World War. In those times and places the undercurrents that they witnessed embraced Cubism, Expressionism, Vorticism, Futurism and trench warfare. For Joseph Wright, the gulf between a soft Derbyshire landscape and an exploding Vesuvius was wide; but nevertheless Wright artfully places a tree or a rock in his depictions of Vesuvius to maintain picturesque convention. For the Icelanders a century and a half later, however, the only reasonable way to convey the enormity of the noise and fire of Grismvotn or Hekla was to call on the reinforcements of twentieth-century modernism. In the 1930s the only metaphor for Grimsvotn was a battlefield cannonade.

Einarsson's Futurist landscape makes a pointed stylistic pairing with *Under the Pergola in Naples* (1914) by Umberto Boccioni (1882–1916). This is Boccioni in a modest, domestic mood, experimenting with Cubism, a world away from his violent Futurist *The City Rises* (1910–11, MOMA, New York). But the analogy is clear: both artists use fragmentation to shatter their images and to create through dislocation a real sense of uncertainty and foreboding. In Boccioni's painting Vesuvius rises above the seated figures and their quiet lunch, the volcano's sharp triangular form being repeated among the faceting across the canvas. While nothing much is happening, the collaged song sheet on the right, with its corner echoing the volcano shape and its cover illustrated by a woodcut of an erupting Vesuvius, is a reminder of the mountain's violent character. The entire broken imagery suggests that things are about to change, as in that instant between the shattering of a mirror and the crash of its fragments to the ground.

Flight from Etna by Renato Guttuso (1912–1997) of 1940 has, like the Boccioni, multiple meanings. In its form and its expression of overpowering violence – natural rather than military – it echoes Picasso's *Guernica* of 1937, and in its political message it is every bit as pointed. Etna erupted in 1928 and flattened the village of Mascali, causing thousands to flee. Mussolini used the event to draw the Sicilian population to his Fascist cause, and

Umberto Boccioni, *Under the Pergola in Naples*, 1914, oil on canvas.

over the following few years his regime rebuilt Mascali with much noise and trumpeting as a monument to the efficiency and zeal of his government. Etna's presence is only lightly suggested in the painting; instead Guttuso expounds the brutal effect of the eruption on the local people, many of whom are left naked and terrified, their horses and cows bolting, their domestic furniture and order scattered. The volcano is here used by Guttuso – a member from 1940 of the Italian Communist Party – as a metaphor for the volcanic effect on Italian society of the Fascist ascendancy.

The painter Gerardo Murillo (1875–1963), who renamed himself 'Dr Atl', having by turns also called himself 'Dr Fox' and 'Dr Orange', had a rich first-hand experience of volcanoes in his native Mexico. 'Dr Atl' was a political as well as an artistic revolutionary, and a volcanologist, social theorist and restless nationalist, art critic, poet and painter.[13] He studied art at the San Carlos Academy in Mexico City, and with a gift from the president of Mexico travelled in about 1900 to Rome, where he was overwhelmed by the power of Italian Renaissance fresco painting, and in particular by Michelangelo's Sistine Chapel ceiling. This trip encouraged him to recognize the political and educational power of public art, and on his return home he galvanized fellow artists including Diego Riviera, José Orozco and David Siqueiros to paint frescoes in Mexico, with the result that Mexican art took a new and powerful direction.

A hugely important figure in Mexican art, Atl was equally fascinated by volcanoes. Abandoning his interest in political painting in the late 1920s, he took on the twin peaks of Mexican symbolism, the volcanoes Popocatépetl and Iztaccihuatl, as his driving pictorial subject-matter: 'I abandoned pictorial pedantry and with fury I began to paint landscapes under realistic criteria',

Renato Guttuso, *The Flight from Etna*, 1940, oil on canvas.

he later wrote.[14] Atl had published his book *Symphonies of Popocatépetl* in 1921, writing of 'the powerful influence throughout my life [of the volcano] . . . From the top of the mountain I have seen the world as a marvellous spectacle without reluctance, profoundly, intensely.' By an extraordinary stroke of good luck and opportunism, he was also able to buy the piece of farmland in which, in February 1943, a volcano had suddenly arisen, spewing ash and lava, and growing within a year to around 300 metres in height. Useless now to the farmer, Atl gave the new volcano – Parícutin, named after its local village – a new role as an artist's model. Dr Atl was a paradox: simple and profound, artist and activist, revolutionary and conservative. 'I was not born a painter,' he said, 'I was born a walker and walking has led me to love nature and the desire to represent it.'

Renato Barisani, *Stromboli*, 1958, mixed media, with lava dust, stone and oil on canvas.

The Neapolitan artist Renato Barisani (*b.* 1918) was in the 1950s a member of the Movimento Arte Concreta of Milan, one of a number of fragile and often short-lived artistic groupings that ebbed and flowed in Italy in the post-war years. He has frequently taken Vesuvius and other Italian volcanoes as his subject-matter, creating images with sand, shells and other objects, which mix with wide fields of colour to form bright and burning, or quietly brooding compositions. His *Stromboli* (1958) is composed of lava dust from the island, with a central knob of volcanic rock, set like a jewel. While it is a curiously sombre piece for so active a volcano as Stromboli, it evokes both a sense of elevation in the viewer, rising high above the island, and the excitement of discovery, as of finding a bright stone or shell on a deserted beach. Thus, the sense of scale is first confused and then abolished.

Writing about the place of emotion in his work, the playwright Eugene Ionesco (1909–1994) used the volcano as a metaphor not for destruction, but for intense creativity:

> For the moment, I exist. Passions slumber in me that might explode, and then be held in check again. Jets of rage or joy lie within me, ready to burst and catch fire. In myself I am energy, fire, lava. I am a volcano. Most often, I am half

> asleep: my craters wait for this continual boiling to rise, emerge, satisfy its instincts; for my incandescent passions to pour out, ignite and spread forth in an assault on the world.[15]

In choosing an active volcano as a subject in their work, artists as different as Gudmundur Einarsson, Renato Guttuso and Dr Atl have touched the quick of human nature. Here is a metaphor that runs deep within the nervous system of humanity to inspire fear and respect – like dragons in the Dark Ages, millenarianism in the eleventh century, the Inquisition during the Counter-Reformation, the nuclear bomb in the twentieth century, and climate change today. But by confronting their volcano, the artists are to some extent controlling the idea of its force, and taking ownership of it. When, in the book of Genesis, Adam named the animals, he was confronting his, and humanity's, fear of the unknown; so the lion became the lion, to do what lions do; the crocodile the crocodile, and so on. It is notable that here these three artists, and many others, chose *their own* volcanoes to address: thus Gudmundur the Icelander paints Grimsvotn; Guttuso the Sicilian paints Etna; and the Mexican Dr Alt presents us with Popocatépetl and Paricutin. At the very least, the uncontrollable is under close observation, and pressures are released.

8 The Shifting Furnace

Volcanic eruptions are continuous and inevitable. As photographs from the International Space Station reveal, on earth at any one time some volcano somewhere is smoking, bubbling or erupting. At the bottom of the Pacific and Atlantic Oceans this process goes on in silence all the time, spawning life forms that have evolved as a response to the continual heat. This planet never rests.

The amazement expressed by those who in 1811 witnessed the birth of the island of Sabrina in the Azores, or in 1831 of Graham island in the Mediterranean, was the same unparalleled wonder that greeted the birth of Surtsey off the south coast of Iceland in 1963. Recognising the fecundity of the earth, 'birth' is the word generally given to the arrival of new volcanoes, from Monte Nuovo near Naples in 1538 and Paricutin in 1943–44, to Sabrina, Graham and Surtsey. Sabrina and Graham were rapidly swallowed up again, but Monte Nuovo, Paricutin and Surtsey – named after Surtr the Norse god of fire – remain and look like becoming fixtures. Birth from the sea is itself a deep-rooted concept, with many echoes across time and art, such as the story of the birth of Venus and the medieval mystery of the origin of Barnacle Geese.[1] But it is nothing new: Iceland itself bubbled up from the volcanic Atlantic Ridge around thirty million years ago, long after the dinosaurs had come and gone from the earth, as did the Azores archipelago and, in the Pacific, all of Hawaii.

Artists are not usually in the first wave of shock troops to arrive at a volcanic event. When the walls of the crater of Mount Nyiragongo in the Democratic Republic of Congo failed in

January 1977, the contents of a red-hot lava lake flowed rapidly down the mountain, overwhelming villages and killing thousands of people. It is impossible to imagine the constant emotional pressures of living on the side of a mountain with such a terrible burden at its peak, like a pot of boiling water lodged just above an opening door. The high speed of the flow down Nyiragongo was because the lava in that volcano has a low silica content and is thus not as sticky as the lava from, for example, Vesuvius, which travels at a far more stately but no less devastating pace. Halut's painting, with the spirits of the volcano rising up within its cloud, takes us directly to the heart of the disaster as the terrible red tongue of fire slips down the hillside and destroys his village. The immediacy of the painting is comparable to Guttuso's *Flight from Etna*, and conveys so much more humanity than the nineteenth-century constructs of Schopin or Poynter.

Given the intense worldwide scrutiny by volcanologists of volcanic 'hotspots', eruptions from the mid-twentieth century have been constantly filmed and assessed even as the first flow of lava appears above the crater's edge. We have immediate aerial film of the proceedings – often with a light plane's wingtip moving evocatively in and out of shot, and the over-excited commentary of a press reporter who is safe and secure in the cockpit. This is of course wholly different from the nineteenth century, when artists were the first accurate witnesses of such events, and from the widespread eighteenth- and nineteenth-century practice of embellishing reports of eruptions for dramatic and narrative purposes. We can seriously doubt Schopin, but we cannot doubt Halut.

Detachment, transformation, terror, imaginative invention and cool analysis: these are the qualities that artists brought in the latter years of the twentieth century to volcanic subjects. Dieter Roth (1930–1998), a German artist who lived in Iceland from 1957, used a well-known aerial photograph taken in 1963 of the birth of Surtsey to make a series of eighteen collotype prints as a demonstration of print-making techniques.[2] Raised up on a platter, and set on a fetching check tablecloth against a sea view, Surtsey is shown to us as a delicious or possibly a

Dieter Roth, *Surtsey*, 1973–4, collotype.

disgusting meal. Roth's art explored food, taste and biodegradation – rot, in short. Associated with the Dadaist group Fluxus in the 1960s, his work provoked, annoyed and fascinated galleries and collectors. He collaborated with Richard Hamilton, Daniel Spoerri, Jean Tinguely, Joseph Beuys and others to make art that was fragile, fleeting and random. His chocolate or cheese exhibits were mostly eaten by gallery goers, or simply rotted away. He painted his photographic portrait of the Swiss collector Carl Laszlo with processed cheese 'to get his goat', and in an exhibit in 1970 he filled 37 suitcases in a Los Angeles gallery with cheese, which after a few days began to stink and crawl with maggots. The whole show was taken out and dumped in the desert. Roth took himself very seriously, as did a wide constituency of galleries, collectors and friends, who hold an annual international conference in his memory. In essence Roth's art is about transience, the consumer society, love, sex and desire, multiplication and variation, and having a good time – in short, it is about life. So Surtsey, arising so suddenly in 1963 from the seabed on Roth's adopted doorstep, became a rich and unexpected metaphor for the transient, being itself a plate of seafood raised up out of the North Atlantic fishing grounds. Will it disappear, be eaten away by the waves, or even quietly rot? Roth, when he made the print series, could not himself be sure, and even now, nor can anybody else.

Roth put himself outside the mainstream, but having sufficient gravitas to draw others to him he was never an 'outsider', despite placing himself in Iceland far from the orthodox Western art axis. He did not seek out his volcano, as did Wright or Church, it just rose out of the sea, like Venus, and presented itself to him. An 'outsider artist' of African-American and Native American descent, Joseph Yoakum (1890-1972) did not begin painting

Joseph Yoakum, *Crater Head, Maui Province, Hawaii National Park*, 1960s, ball point pen and watercolour.

and drawing until he was in his sixties. His landscapes writhe and twirl under his dancing ballpoint pen, and are drawn from his memory of places he had visited during his long life. Among these were some of the volcanoes on Hawaii: *Crater Head, Maui Provience [sic], Hawaii National Park* shows the volcano like a pile of cream cakes, melting over one another in a manner not unlike the cursive quality of some of Dieter Roth's drawings. Certainly, Yoakum presents his volcano with the same delicious affection with which Roth presents Surtsey.

A spectacular volcanic event took place in the United States in the weeks before 18 May 1980 when, after a series of earthquakes of gradually increasing magnitude, the side of Mount St Helens in Washington State blew off, creating devastation in this sparsely inhabited part of the American wilderness. It would be meaningless to quote the immense figures of the volume of mud and lava spilled out (2.3 cubic km), or the strength of the blast (27,000 times stronger than the atomic bomb dropped on Hiroshima), or the length of roads destroyed (480 km), or the cost of the damage ($10,000 million) or the number of deaths (57 people), or that the landslide, one of the largest in recorded history, travelled at 240 km per hour (150 mph). Rather, we can leave it to the British artist Michael Sandle (*b.* 1936) whose series of four large watercolour and ink drawings record the event, which happened to take place on his 44th birthday.

Made from a series of aerial photographs, these works are breathtaking in scale and substance, having the immediacy of a series of freeze-frame film images that momentarily hold back the inevitable sequence of the explosion. On the left of each sheet is a moment from the explosion itself; on the right an abstracted obliteration of each moment, cancelled by either a grey abstracted blankness or a determined black diagonal stroke.[3] These are part of a group of works that Sandle made in the 1980s that explore themes of devastation: a large diptych of 1984 is an image of the meltdown of the Chernobyl nuclear reactor, while other drawings and sculpture take the disasters and memorials of twentieth century warfare as their theme. A volcanic subject is unique in Sandle's art, but treating as he does with the

Michael Sandle, *Mount St Helens, Version III and IV*, 1981–2, watercolour.

result of overpowering force the *Mount St Helens* series is in step with his melancholy bronze catafalques such as *The End of the Third Reich* (1980) and his violent *A Mighty Blow for Freedom (Fuck the Media)* (1988).

A few years after Surtsey had appeared out of the sea off Iceland, another significant eruption occurred on the island of Heimaey in the Vestmannaeyjar archipelago of which Surtsey had become part. In January 1973 an earthquake shower shook the island, and fissures began to run across it, spewing out ash, steam and lava. This deteriorated and within days a new mountain – later named Eldfell (Mountain of Fire) – began to form, rising to about 200 metres after three weeks. Keith Grant (*b.* 1930), an artist who had always been attracted to the pictorial allure of extreme environments, flew over Heimaey in 1973 when the column of smoke and ash was still ascending. At 6,000 metres (20,000 feet) he found a perspective on the eruption that prefigures that taken by Michael Sandle's source photographs, and one that is unique to the twentieth century, when aerial viewpoints became possible. This would not be a perspective likely to be taken advantage of by any nineteenth-century artist, raised aloft in a gas balloon. Grant's starkly vertical and compellingly beautiful eruption column, echoing like Gudmundur Einarsson's *Grimsvotn* the form of a nuclear mushroom cloud, is the central feature of the painting. It echoes also the articulation of the human figure, the lower half of the body running out up and away from the crater.

Keith Grant has made a group of paintings of volcanoes, most of which originate from Iceland. The crisp cold of the far north – its emptiness, the stark crystalline forms, the limpid blues and wide even limitless horizons – attracted Grant away from tranquil temperate landscapes. Seen in contrast with his paintings of equatorial subjects in Africa and South America that pulse with colour and heat, Grant's Iceland volcanoes project a sense of control and chill detachment that counteracts the chaos and disorder of their subject matter.

Ilana Halperin (*b.* 1973) noticed, when she turned 30 in 2003, that Eldfell too was 30 years old. To mark this joint birthday

Ilana Halperin, *Emergent Landmass*, 2006, etching.

she visited Eldfell, subsequently creating an exhibition *Nomadic Landmass*, which was shown at Doggerfisher Gallery in Edinburgh in 2005. Halperin has written:

> My work explores the relationship between geological phenomena and daily life. Whether boiling milk in a 100 degree Celsius sulphur spring in the crater of an active volcano or celebrating my birthday with a landmass of the same age, the geologic history and environmental situation specific to the locale directly informs the direction each piece takes.[4]

Nomadic Landmass included photographic images taken from the air over Eldfell, and geological specimens and drawings taken from photographs of the destruction caused by the birth of the mountain. A further project of a volcanic nature, *Emergent Landmass* (2006) explored the short life of Graham Island (Ferdinandea), appearing and disappearing in the Mediterranean in 1831. Halperin has taken the extreme detachment of volcanic activity as her subject, and has personalized it, drawn it to herself, and invited it to become intertwined with her own life. The mountain's pulse, and hers, become one.

Keith Grant, *Eruption Column 20,000 Feet, Heimay, Iceland,* 1973, oil on board.

James P. Graham, still from *Iddu*, 2007.

The filmmaker James P. Graham has for some years been working on the relatively more durable Mediterranean island, Stromboli, filming its volcanic activity with a circle of twelve cameras. With these he has created and developed his work *Iddu* (2002–7): a word that means 'him' in Sicilian dialect. With *Iddu*, the viewer is an active participant, enveloped by the 360° event, only too aware that the natural forces that sustain life on earth could also destroy it. To the straining ventral harmonies of Stromboli, Graham reveals the complex relationship on the volcanic island between the constant sea, the conic fluxing landmass and the effusive lava that creates it. Rhyming shapes populate his film: for example, the pyramidal island is echoed in form by a shower head adapted to drip nectar to gathering bees. The nectar, shown at the conclusion of the piece, is a metaphor for the fertility of the emerging lava, while rolling down the sides of the island fiery lumps of this same lava dance and spin like a tumbler in a circus.

The global financial crisis of October 2009 prompted Ian Brown to extend his existing series of prints exploring natural disasters into volcanic themes. Using his collection of volcano postcards as a starting point he began a body of work that could stand as a metaphor for the situation at the time. Referring to the large scale of his prints, 152 x 107 cm, he has written: 'It is interesting to note the psychological difference scale makes. The viewer tends to look *at* prints of a conventional size, standing outside, looking in, whereas confronted with an image almost body-size and filling the field of vision, the possibility of entering and inhabiting it occurs.' Brown noticed a subtle dialogue

between source photograph, postcard and screenprint: 'The photographs are a statement of fact, the hand coloured version a poetic interpretation, and the print underlines the illusion involved in the transcription of the three-dimensional world into a two-dimensional print.'[5]

Eleanor Antin (*b.* 1935) is a filmmaker and performance artist based in California. Her sequence of photographs *The Last Days of Pompeii* (2001) is, as the artist has said in a recorded interview, deliberately based on nineteenth-century academic painting, 'because we are dealing with England and France, who invented Rome in a way.'

> They invented Rome to help glorify their own role as great colonial powers with India and Africa and whatever they were colonizing. And especially the Brits. They had an enormous fascination with Roman subjects and obviously did see themselves in that guise as the new Rome. So I thought that it would be interesting to see our relationship through the eyes of nineteenth century salon painting, which I've always loved anyway. It's such campy painting . . .
>
> I also wanted to have someone's eyes, a single character, who would be a Pre-Raphaelite woman who would look and watch and observe everything in her wheelchair. She's in a wheelchair and only in the end and final picture – which is the only one that deals with the devastation and the destruction – is she standing up. Once the disaster happens she stands and walks.[6]

Eleanor Antin, 'A Hot Afternoon', from *The Last Days of Pompeii*, 2001, chromogenic print.

The photographs were staged at the villa La Jolla in Rancho Santa Fe, California, chosen as the site for the recreated Pompeii because of its dramatic location:

> The town was laid out on this incredible bay – like Pompeii – which isn't exactly on a bay, but it's filled with these beautiful affluent people living the good life on the brink of annihilation. And in California we are slipping into the sea, eroding into the sea, living on earthquake faults. And we finished this piece two-and-a-half weeks before 9/11. So the relationships between America as this great colonial power and Rome – one of the early, great colonial powers – were extremely clear to me. And I think it's pretty much clear to everybody from the work.

Ian Brown, *Natural Disaster: Variant I*, 2009, screenprint on tub-sized Somerset Satin paper.

There is a Hollywood look to Antin's photographs, and this inevitably leads them to being read as film stills. However, like the nineteenth-century paintings that they emulate, they are detached moments in the story of the days before the cataclysm, and as such are visually and literally direct heirs to paintings like Schopin's *Last Day of Pompeii* and Poynter's *Faithful unto Death*. The only thing that distinguishes them from the nineteenth-century works is that they are photographs and the others are oil paint on canvas. Each is an imaginative and as far as possible faithful interpretation of their shared subject, and each within the artistic practices of the time, used hired people as models. The coincidence that the piece was finished only two and a half weeks before 9/11 gives added poignancy and an unwanted further layer of meaning. In the very same way that we cannot now look at pictures of the Twin Towers without remembering 9/11, so we cannot look at images entitled *The Last Days of Pompeii* without recalling the eruption of Vesuvius.

Georgia Papageorge, *Kilimanjaro – Southern Glaciers 1898*, mixed media on canvas with inkjet print taken from the earliest known photograph of the glaciers, and poured ash from Kilimanjaro, 2010.

Decades before Eleanor Antin used Hollywood as an inspiration for her *Last Days of Pompeii*, Hollywood let itself loose on the subject of volcanoes, big time. What better a theme for a disaster movie than a volcano? Hollywood has given us a towering inferno and a disoriented Great White Shark, and so true to form we have also had *Krakatoa – East of Java* (dir. Bernard K. Kowalski, 1969) in wrap-around Cinerama, *Volcano* (dir. Mick Jackson, 1997), featuring a volcano erupting under Los Angeles, *Disaster Zone: Volcano in New York* (dir. Robert Lee, 2006), in which New York gets a visit from the wandering lava, and *2012* (dir. Roland Emmerich, 2009), the year the world faces the eruption of the Yellowstone caldera. There have been six movie versions of *The Last Days of Pompeii*, three of them Italian, three American, and no doubt there are more to come. Geographically sensitive readers will have noticed that Krakatoa is actually west of Java, but when has an erroneous compass point or a misspelling ever hampered a film publicist's career? Moving swiftly to opera, Patrick Leigh Fermor's novel was the source of Malcolm Williamson's opera *The Violins of Saint-Jacques*, produced at Sadler's Wells in London in 1966.

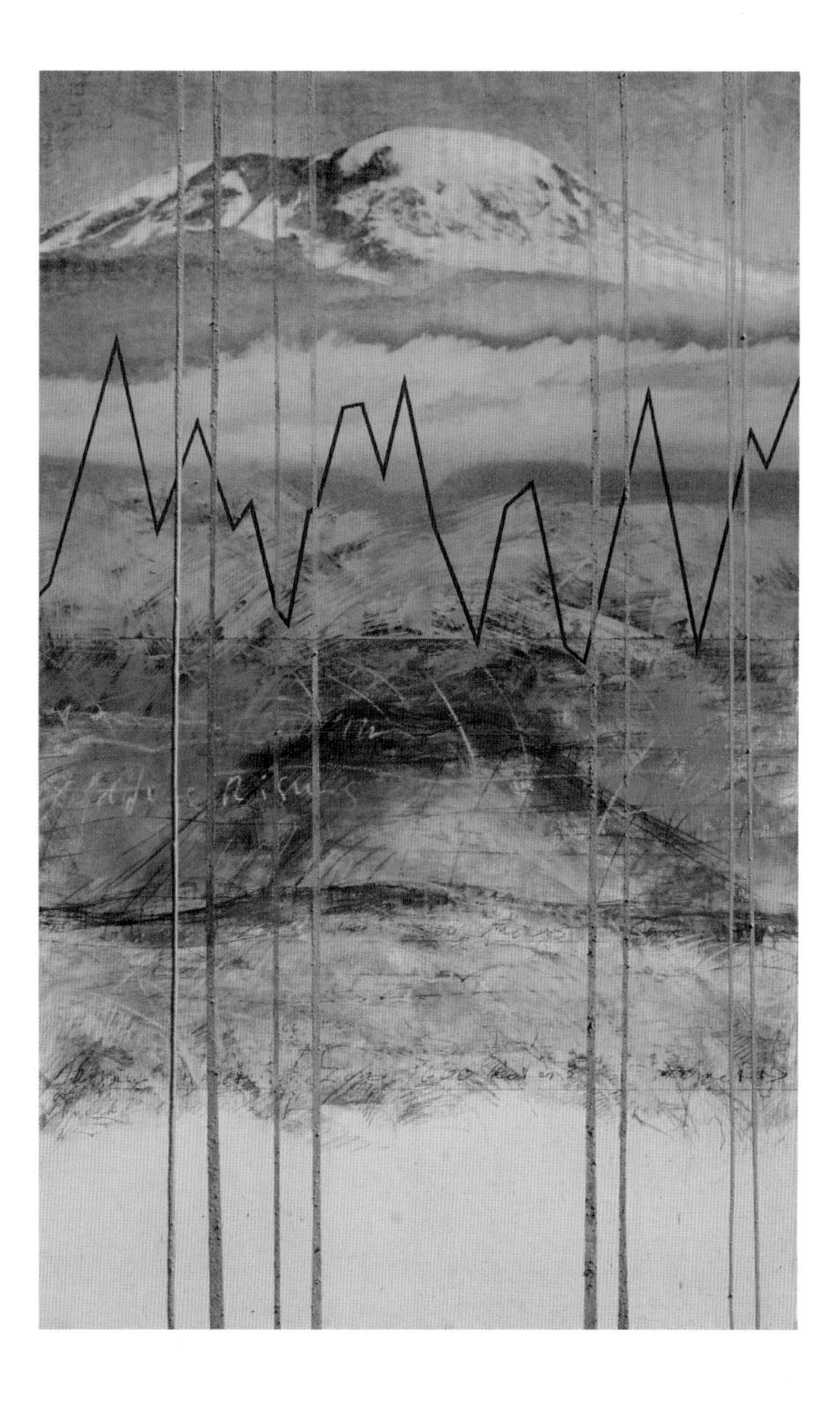

The role of the artist is to follow, or to create, an iconography of the subject in hand. Georgia Papageorge (*b.* 1941) is among the first artists to begin the task of creating an iconography for Kilimanjaro, the dormant volcano in Tanzania. Her palette consists of paint and canvas, photographs, charcoal, tree bark, red and white chevron barrier cloth and the fertile product of the volcano's own interior, lava dust. The earliest known photograph of the mountain, taken in 1898, is enlarged by her and streaked with trails of liquid lava, and articulated by a red zigzag line representing temperature fluctuations and glacier melt on the volcano over the twentieth century. Other works deploy sacks of charcoal, the 'black gold' made from burning the hardwood of the surrounding plains, thus destroying for small local gain the very same trees that resist the process of desertification and global warming. Papageorge warns us of the ruination being wrought on the landscape and thus on the climate by uncontrolled cutting and burning.

Growing knowledge of the origin and causes of volcanic eruptions gives scientists the ability to predict them and warn of their imminence. Earth tremors are the harbingers of eruptions, and to measure them systems of sensors are installed by the World Organization of Volcano Observatories (WOVO) on the sides of active volcanoes to measure earth movement. While eruptions can therefore be more or less effectively predicted close to the event, they cannot be stopped or averted. We are living by mutual consent on top of the furnace that keeps us alive, and from time to time the furnace shifts and resets itself through the cracks in the shell that keeps its heat in. The myth of Typhon, scratching himself, and tossing and turning beneath Etna, is not, as it turns out, all that far from the truth. A report published by the Geological Society of London in 2005 discussed the global effects and future threats of super-eruptions, those massive volcanic events that disrupt the regular cycle of the planet.[7] They have occurred before, at Jemez Mountains, New Mexico (1.1 million years ago); at Toba, Sumatra (74,000 years ago); at the Phlegraean Fields outside Naples (35,000 years ago); and at Kikai, Japan (6000 years ago) – to name only four, in increasing

Eruption of Mount Etna,
Sicily, April 1983.

proximity of time. Living as we do on top of this shifting furnace, we should remind ourselves that they will occur again. As the Geological Society report asserted, 'It's not a question of "if" – it's a question of "when"'. The result of ash clouds and associated gases flung into the atmosphere by a super-eruption would be global cooling by up to 5°C – sufficient to cause a new Ice Age, and freeze and kill the equatorial rainforests.

But Andy Warhol (1928–1987) was not thinking about all that when he painted his *Vesuvius* (1985) and made the many prints of this same exuberantly violent image. He takes us straight back to the late eighteenth century, particularly to Joseph Wright, in this powerfully blasting image that looks as much like an upended jet engine as it does Vesuvius. The acrylic paint, applied carefully within the clearly demarcated areas, has nevertheless produced lava of its own in the unexpected paint spatters that fall down the side of the canvas. As well as the general tenor of the

Andy Warhol, *Vesuvius*, 1985, acrylic on canvas.

Mud bath crater at the Tizdar volcano in the Sea of Azov, Krasnodar Krai, Russia, 2009.

composition, Warhol has another thing in common with Wright: like Wright he produced multiples of this image, but he differed from the British artist in varying the colour balance of his volcanoes, producing lava that is sometimes yellow, sometimes, red, sometimes black.

The paintings by the New York artist David Clarkson take the natural detachment of volcanoes to an extreme, and use the moving image in a manner quite at odds with the way James P. Graham uses it. Clarkson's sources are the CCTV images of volcanoes in their living state, but here flattened in substance, and with diminished colour that distances them emotionally from the viewer. Further distancing comes from the inclusion of computer rubric giving date, time and place at the edges of some of the works.

We have travelled great distances, artistically and emotionally, from Pierre-Jacques Volaire to David Clarkson, and it is a reasonable guess that we have further yet to go. With increasing awareness of the power of volcanoes to disrupt us at the very least, and destroy us at the worst, artists have a new weapon. However powerfully or inexplicably they depict volcanoes as their subject, volcanoes will themselves surprise even the most inventive and imaginative of artists. It may be that current political, social and economic structures are too finely tuned to survive unscathed a volcanic eruption on the scale of Laki (1783), Tambora (1815) or Krakatoa (1883). The volcano, on this planet at least, will probably have the last hurrah.

Tangerines covered in volcanic ash at Takara township in Miyazaki prefecture, after the eruption of Shinmoe-dake, a volcano in southern Japan.

ERUPTION TIMELINE

This timeline covers volcanoes discussed in this book whose eruptions have touched on myth, reportage, literature and art.

6200BC	Karapinar, Anatolia (modern-day Turkey)
6000	Mazama, Oregon, USA
1620	Santorini, Aegean Sea
479	Etna, Sicily
400	Vulcano, Mediterranean Sea
49	Etna, Sicily
AD 38–40	Etna
79	Vesuvius, Italy
580	Vesuvius
1104	Hekla, Iceland
1341	Hekla
1538	Monte Nuovo, Naples, Italy
1600	Huaynaputina, Peru
1631	Vesuvius
1669	Etna
1707	Vesuvius
1730–37	Vesuvius
1754	Vesuvius
1760–67	Vesuvius
1770s	Vesuvius
1783–4	Laki, Iceland
1784	Misti, Peru
1803	Cotopaxi, Ecuador
1805	Vesuvius
1808	Agung, Bali
1811	San Miguel, Azores
1812	Souffrière, St Vincent, West Indies
1814–15	Mayon, Philippines; Vesuvius

1815	Mount Tambora, Philippines
1817	Raung, Java, Indonesia
1818	Colima, Mexico
1822	Usu, Japan; Galunggung, Java; Vesuvius
1823	Katla, Grimsvotn, Iceland
1826	Vesuvius
1831	Graham Island, or Ferdinandea, Mediterranean Sea
1839	Vesuvius
1841	Erebus, Antarctica
1850–58	Vesuvius
1850–63	Cotopaxi, Ecuador
1857	Sangay, Ecuador
1863	Mauna Loa, Hawaii
1881	Kilauea, Hawaii
1878	Hekla
1879–80	Ilopango, El Salvador
1883	Krakatua, Java
1886	Tarawera, New Zealand
1902	Pelée, Martinique, West Indies
1902–3	Grimsvotn, Iceland
1906	Vesuvius
1913	Hekla
1928	Etna
1934	Grimsvotn
1943	Paricutin, Mexico
1944	Vesuvius
1963	Surtsey, Iceland
1973	Eldfell, Heimay, Iceland
1977	Nyiragongo, Democratic Republic of Congo
1980	Mount St Helens, Washington state, USA
1991	Unzen, Japan
2002	Nyiragongo
2006	Fourpeaked Mountain, Alaska
2010	Eyjafjallajökull, Iceland

REFERENCES

1 'The Whole Sea Boiled and Blazed'

1 Ian Hodder, *Catalhöyük: The Leopard's Tale* (London, 2006), pp. 162–3. An alternative view holds that the wall painting does not represent a volcano or the town, but a leopard skin beside a geometric design. Stephenie Meece, 'A Bird's Eye View – Of a Leopard's Spots: The Çatalhöyük "Map" and the Development of Cartographic Representation in Prehistory', *Anatolian Studies*, 56 (2006), pp. 1–16.
2 Hodder, *Catalhöyük*, pp. 174–5.
3 Thucydides, *The Peloponnesian War*, book 3, 88, trans. Rex Warner (Harmondsworth, 1966), p. 213.
4 *Odyssey*, Book 9. Homer describes the Cyclopes as being shepherds and farmers, not blacksmiths.
5 Aeschylus, *Prometheus Bound*, trans. A. J. Podlecki (Oxford, 2005), p. 101, ll. 364–5.
6 Virgil, *The Aeneid*, book III, ll. 571–82, trans. W. F. Jackson Knight (Harmondsworth, 1966), p. 92.
7 Plato, *Phaedo*, 112e–113a, trans. David Gallop (Oxford, 1975), p. 67.
8 Virgil, *The Aeneid*, book VI, l. 548ff., trans. W. F. Jackson Knight (Harmondsworth, 1966), p. 163.
9 Pindar, *Pythian Ode*, I, 15–24, from William H. Race, ed. and trans., *Pindar: Olympian Odes, Pythian Odes* (Cambridge, MA, 1997), p. 215.
10 *Meteorologica*, 367a.
11 *The Geography of Strabo*, book 1, ch. 3, paras 9–10, trans. H. L. Jones (Cambridge, MA, 1931), vol. I, p. 199. But see also 6.1.6.
12 Ibid., Book 1, ch. 3, paras 15–16, p. 213.
13 Augustine, *City of God*, book 20, ch. 8, cf. Revelation, 20:9–10.

2 The Geography, Science and Allure of Volcanoes

1 Plato, *Phaedo*, 110b, trans. R. Hackforth (Cambridge, 1955), p. 176.

2 Lynne Green, *W. Barns-Graham: A Studio Life* (London, 2001), p. 246.

3 Norman Lewis, *Naples '44* (Glasgow, 1978), p. 104; entry for 22 March.

4 Ilana Halperin, *Field Diary, Hawaii* (2008), www.ilanahalperin.com; personal communication.

3 'A Horrid Inundation of Fire'

1 Pliny the Elder, *Natural History*, book II, 236 and III, 92, in Pliny the Elder, *Natural History – A Selection*, trans. John F. Henley (Harmondsworth, 1991).

2 Pliny the Younger to Cornelius Tacitus, in *The Letters of the Younger Pliny*, ed. Betty Radice (Harmondsworth, 1963), p. 166.

3 Ibid., p. 172.

4 Strabo, *The Geography of Strabo*, book 5, ch. 4, trans. H. L. Jones (Cambridge, MA, 1931), vol. I, para 8.

5 Alwyn Scarth and Jean-Claude Tanguy, *Volcanoes of Europe* (Fairfax, VA, 2001), pp. 51–69.

6 Benedeit, 'St Brendan's Voyage', ch. 24, ll. 1420–28.

7 *Flatey Book* ['Flat Island Book']. Sigurdur Thorarinsson, *Hekla: A Notorious Volcano* (Reykjavík, 1978).

8 Uno von Troil, *Letters on Iceland, containing Observations . . . made during a Voyage undertaken in the year 1772 by Joseph Banks . . . Dr Solander, Dr J. Lind, Dr Uno von Troil . . .* (London, 1780), p. 226. The first recorded eruption in Iceland (in the ninth century) produced a stretch of lava 14 km (9 miles) long by 4 km (2½ miles) wide, ibid., p. 225.

9 I am most grateful to Professor Oswyn Murray for his translation of the lines by Neckam. The Latin is as follows:

Si tamen esse nigram terram, tria caetera censes
Candida, non deerunt qui tueantur idem.
Sed quid? Vulcano dat visus in esse ruborem,
Respondent, sed quid afficit ille rubor?
Materiam flammamque potens connectit in unum
Naturae virtus, rebus amica comes.

Alexander Neckam, *De Laudibus Divinae Sapientiae*, 4, ll. 49–54, from *De Naturis Rerum and De Laudibus Divinae Sapientiae: Chronicles and Memorials of Great Britain and Ireland during the Middle Ages*, ed. Thomas Wright (London, 1863), p. 421.

10 Henry Oldenburg, secretary of the Royal Society, to Thomas Harpur at Aleppo, north Syria, 22 May 1668; A. R. Hall and M. B.

Hall, eds, *The Correspondence of Henry Oldenburg*, vol. IV: *1667–1668* (Madison, WI, 1967). These queries were posed at the Royal Society meeting on 7 May 1668. The original letter in Latin can be found at the Royal Society, EL/01/67.

11 Joscelyn Godwin, *Athanasius Kircher's Theatre of the World* (London, 2009).

12 Athanasius Kircher, *The VULCANO's: or, Burning and Fire-Vomiting MOUNTAINS, Famous in the World: With their REMARKABLES. Collected for the most part out of KIRCHER's Subterraneous World; And expos'd to more general view in English, upon the Relation of the late Wonderful and Prodigious Eruptions of ÆTNA. Thereby to occasion greater admirations of the Wonders of Nature (and of the God of Nature) in the mighty Element of Fire* (London, 1669).

13 Ibid., pp. 6–7.

14 Ibid., 'Explication of Schemes, out of Kircher', preface.

15 Ibid., p. 35.

16 Silvester Darius to King James V of Scotland, Rome, 24 October 1538; British Library, London, Royal MS 18 B. VI, ff. 53v–54r.

17 Scarth and Tanguy, *Volcanoes of Europe*, p. 65.

18 William Bray, ed., *The Diary of John Evelyn* (London, 1907, revd 1952), vol. I, pp. 152–3.

19 Francesco Serao, *The Natural History of Mount Vesuvius*, trans. from the Italian (London, 1743), pp. 40–41.

20 Ibid.

21 Ibid.

22 Anonymous letter from Naples, 10 July 1737; British Library, London, Sloane MS 2039, ff. 44–50.

23 Patrick Brydone, *A Tour through Sicily and Malta* (Dublin, 1780), p. 132.

24 Volcanic Explosive Index (VEI) scale 2, indicating an 'explosive' event in which more than a million cubic metres of material is ejected.

25 'Exploration of the volcano Misti, Near Arequipa', British Library, London, Add. MS 30170B.

26 Entry on 'John Warltire', *Oxford Dictionary of National Biography* (Oxford, 2004); N. G. Coley, 'John Warltire, 1738/9–1810, Itinerant Lecturer and Chemist', *West Midlands Studies*, 3 (1969), pp. 31–44.

27 Royal Society, London, CB/1/3/182.

28 Royal Society, Classified Papers/4ii /11; 'Description of a burning volcano in the County of Kerry ... as communicated by letter to the Rt. Rev. the Bishop of Kenmore', *Dublin Evening Post*, 1/84, 24–7 March 1733.

29 Joseph Priestley to Josiah Wedgwood, 18 May 1781, Royal Society, MM/5/6.

4 Sir William Hamilton and the Lure of Vesuvius

1 Sir William Hamilton to Sir Joseph Banks, 14 June 1801, inserted in vol. I of the Diary of Father Antonio Piaggio, Royal Society, MS/2–9.

2 William Hamilton, *Campi Phlegraei* (Naples, 1776), vol. I, p. 5.

3 Ibid. Five letters to the Royal Society: 10 June 1776, 'Observations on Mount Vesuvius etc.', to Earl of Morton, President of Royal Society, pp. 14–20; Postscript, 3 February 1767, pp. 20–21; 'a plain narrative of what came immediately under my observation, during the late violent eruption, which began October 19 1767 ...', 29 December 1767, to the Earl of Morton, pp. 22–32; 'On the landscape and the inhabitants following the 1767 eruption', 4 October 1768, to Matthew Maty, Secretary of Royal Society, pp. 33–6; 'An account of a journey to Mount Etna', 17 October 1769, to Matthew Maty, pp. 37–52; 'Remarks upon the Nature of the Soil of Naples, and its Neighbourhood', 16 October 1770, to Matthew Maty, pp. 53–89. Hamilton, *Campi Phlegraei*, vol. II, map and 54 plates after Peter Fabris, engraved and coloured.
Supplement: 'Account of the great eruption of Vesuvius ... August 1779'. Dated Naples 1779.
Hamilton, *Campi Phlegraei*, vol. III, letter to Joseph Banks, President of the Royal Society, London, dated Naples, 1 October 1779, pp. 1–28, with five plates by Peter Fabris.

4 Hamilton, *Campi Phlegraei*, vol. I, p. 3, letter to Sir John Pringle, President of the Royal Society of London, dated Naples, 1 May 1776.

5 Ibid., p. 18, letter 10 June 1766.

6 Anonymous letter, Naples, 13 August 1805; BL, Egerton MS 3700, B. Barrett Collection, vol. XII, ff. 110–13b.

7 Hamilton, *Campi Phlegraei*, vol. II, pl. 38.

8 Emilie Beck-Saiello, *Le Chevalier Volaire: Un Peintre Français à Naples au XVIIIe Siècle*, Centre Jean Bérard and Intitut Français de Naples, 2004, fig. 5. See especially pp. 21 and 35.

9 Ian Jenkins and Kim Sloan, *Vases and Volcanoes: Sir William Hamilton and his Collection*, exh. cat., British Museum, London (1996), p. 114.

10 'Memoirs of Thomas Jones', 13 September 1778, *Walpole Society*, XXXII (1946–8), p. 77.

11 Joseph Wright to Richard Wright, 11 November 1774, transcribed in Elizabeth E. Barker, 'Documents relating to Joseph Wright "of Derby" (1734–97) – Letters', *Walpole Society*, LXXI (2008/9), p. 84.

12 Judy Egerton, *Wright of Derby*, exh. cat., Tate, London (1990), p. 166, cat. 101.

13 Joseph Wright to William Hayly, 31 August 1783. Quoted ibid., p. 11.

14 Egerton, *Wright of Derby*, p. 178, cat. 108. It is unlikely that Wright travelled so far south as Sicily; his *View of Catania* was probably painted from some other artist's sketch and indeed may not depict Catania.
15 'Memoirs of Thomas Jones'.
16 Ibid., p. 78.
17 Ibid., pp. 79–80.
18 Quoted Judy Egerton, *Wright of Derby*, cat. 67, pp. 129–30.
19 Faraday's Continental Diary, IET archives, published as Brian Bowers and Lenore Symons, eds, *'Curiosity Perfectly Satisfied': Faraday's Travels in Europe, 1813–1815* (Stevenage, 1991), p. 108.
20 Ibid., p. 161.
21 Mary Somerville to Mrs Samuel Somerville, Naples, 1 March 1818. Dep. c360, MSFP55, Somerville Papers, Somerville College, held by the Bodleian Library, University of Oxford.
22 Cross-written letter, Mary Somerville to Revd Dr Somerville, Naples, 24 March 1818. Dep. c360, MSFP 44. Somerville Papers, Somerville College, held by Bodleian Library, University of Oxford.
23 Mrs Anna Jameson, *A Lady's Diary* (London, 1826).
24 Alwyn Scarth and Jean-Claude Tanguy, *Volcanoes of Europe* (Fairfax, VA, 2001), p. 171.
25 Gilbert White, *Natural History and Antiquities of Selborne* (London, 1789), Letter LXV to Daines Barrington, 25 June 1787.
26 Haraldur Sigurdsson, ed., *Encyclopedia of Volcanoes* (San Diego, CA, 2000), pp. 10–11.
27 Lord Byron, *Childe Harold's Pilgrimage* (London, 1816), canto III, v. 28.
28 J.M.W. Turner to James Holworthy, 11 September 1816, in John Gage, ed., *Collected Correspondence of J.M.W. Turner* (Oxford, 1980), no. 68.
29 Thomas Cayley to his sister Mary, from St Vincent, 30 August 1812. British Library, Add. MS 79532 C/1. See also report in *Philosophical Magazine*, XL (July–December 1812), pp. 66–71.
30 'Report of a Committee of the House of Commons containing an account of the eruption of the Souffrière Mountain in St Vincent's', given to the Geological Society on 21 May 1813.
31 Sam Smiles, 'Turner and the Slave Trade: Speculation and Representation 1805–40', *British Art Journal*, VII/3 (Winter 2007/8), pp. 47–54.
32 Diary of Hugh Perry Keane, Virginia Historical Society, VHS M 551 1 K 197a, 3–30.
33 Pencil sketches of the Bay of Naples, apparently drawn on the flanks of Vesuvius, and studies of the approach to the crater, are in Tate, London, Turner Bequest CLXXXIV, 41a–44a.
34 S. Tillard, 'A Narrative of the Eruption of a Volcano ... off the

Island of St Michael', *Philosophical Transactions of the Royal Society*, CII (1812), p. 152.

35 Lord Byron to John Murray, 8 May 1820, in *Byron's Letters and Journals*, ed. Leslie A. Marchand, 11 vols (London, 1973–82), vol. VII, p. 98.

36 James Hamilton, *London Lights – The Minds that Moved the City that Shook the World, 1805–1851* (London, 2007), pp. 176ff.

37 Ambrose Poynter to Robert Finch, London, 26 April 1822. Finch Papers, d.13, f. 435–6, Bodleian Library, University of Oxford.

38 Sir Charles Morell (aka James Ridley), 'Sadak and Kalasrade', in *Tales of the Genii* (London, 1791).

39 Uno von Troil, *Letters on Iceland, containing Observations . . . made during a Voyage undertaken in the year 1772 by Joseph Banks . . . Dr Solander, Dr J. Lind, Dr Uno von Troil . . .* (London, 1780), p. 231.

40 Ibid., p. 233.

41 M. Dorothy George, *Catalogue of Political and Personal Satires in the British Museum*, 11 vols (Oxford, 1870–1954), vol. VII, p. 99, no. 8479.

42 Ibid., vol. VIII, p. 38, no. 9753.

43 Ibid., vol. VI, pp. 792–3, no. 7863.

44 R. E. Raspe, *The Surprising Travels and Adventures of Baron Munchausen* (new edn, London, 1819), pp. 69ff.

5 The First Days of Graham Island and the Last Days of Pompeii

1 'An Account of the Phenomena of the New Volcano which Rose from the Sea between the Coast of Sicily and the Island of Pantelleria, July 1831. Read in the Hall of the Royal University of Students in Catania by Dr C[arlo] Gemmellaro, Aug 28th 1831', British Library, London, Stowe 1072.

2 John Davy, *Notebook*, front cover inscribed *New Volcano / 1831*. Royal Institution of Great Britain, JD/47.

3 Carlo Gemmellaro, 'An Account of the Phenomena of the New Volcano', p. 8.

4 *The Times*, 23 September 1831.

5 Carlo Gemmellaro, 'An Account of the Phenomena of the New Volcano', p. 6.

6 John Davy, Notebook, Royal Institution of Great Britain, JD/47. See also J. Davy, 'Some Account of a New Volcano in the Mediterranean . . .', *Proceedings of the Royal Society of London*, CXXII (1832–3), pp. 237–49, 251–2; CXXIII, pp. 143–5. Also J. Davy, 'Some Remarks in Reply to Dr Daubney's note . . . over the site of the recent volcano in the Mediterranean', *Philosophical Transactions of the Royal Society*, CXXIV (1834), Part I, p. 552.

7 *The Times*, 14 September 1831.
8 W.E.K. Anderson, *The Journal of Sir Walter Scott* (Oxford, 1972), p. 682, n.3.
9 J. G. Lockhart, *Memoirs of Sir Walter Scott* (Edinburgh 1882), vol. x, pp. 126ff.
10 F. Beaufort, Admiralty, to John Davy, 22 March 1842, in John Davy Papers, book of sketches and notes, 1831–4, Royal Institution, JD1/4/1.
11 Full title: *Geological Observations on Volcanic Islands, with Brief Notices on the Geology of Australia and the Cape of Good Hope.*
12 Charles Darwin to John Henslow, March 1834, available online at www.darwinproject.ac.uk/darwinletters/calendar/entry-238.html
13 'George Poulett Scrope', *Oxford Dictionary of National Biography* (Oxford, 2004).
14 'Sir Charles Lyell', *Oxford Dictionary of National Biography.*
15 *American Quarterly Review*, XVI/507, quoted in 'Edward Bulwer Lytton', *Oxford Dictionary of National Biography.*
16 Leslie Mitchell, *Bulwer Lytton: The Rise and Fall of a Man of Letters* (London, 2003), p. 40.
17 An account of the place of Vesuvius in literature and drama is Nicholas Daly, 'The volcanic disaster narrative: from pleasure gardens to canvas, page and stage', *Victorian Studies*, LIII/2 (Winter 2011), pp. 255–85.
18 W. G. Carey, trans., *The Complete Works of Alexander Pushkin*, vol. III: *Lyric Poems: 1826–1836* (Horncastle, Lincs, 1999), p. 189.
19 The English translation of Pliny's letters that Bulwer Lytton would have known was first published in 1746. The translation used here in chapter Three is by Betty Radice (Harmondsworth, 1963).
20 Edward Bulwer Lytton, *Last Days of Pompeii* (London, 1834), book 5, chap. 4.
21 Ibid, book 5, chap. 9.
22 Ibid, book 5, chap. 7.
23 Arlene Jacobowitz, *James Hamilton 1819–1878 – American Marine Painter*, exh. cat., Brooklyn Museum, New York (1966), esp. p. 9; cat. 36.
24 This point was first made by Teresa Carbone in the entry on James Hamilton in Teresa A. Carbone, ed., *American Paintings in the Brooklyn Museum: Artists Born by 1876* (London and New York, 2006) vol. II, pp. 586-92. I am grateful to Karen Sherry for bringing this to my attention.
25 *Godey's Magazine and Lady's Book* (1847). Quoted Carbone, ed., *American Paintings in the Brooklyn Museum.*
26 Quoted Edward Morris, *Victorian and Edwardian Paintings at the Walker Art Gallery and at Sudley House*, National Museums on

Merseyside, exh. cat. (1996), p. 363–6.
27 Ibid.; and Bulwer Lytton, *Last Days of Pompeii*, book 5, chap. 6.
28 William Rookes Crompton to Henrietta Matilda Crompton, Naples, 11 March 1818, North Yorkshire County Record Office, MIC 2728/398; M. Y. Ashcroft, ed., *Letters and Papers of Henrietta Matilda Crompton and her Family*, North Yorkshire County Record Office Publications, no. 53 (1994), p. 41.
29 J. C. Dahl's diary, Oslo, University Library, MS 80 1001. Quoted *'Nature's Way' – Romantic Landscapes from Norway*, ed. Jane Munro, exh. cat., Whitworth Art Gallery, Manchester, and Fitzwilliam Museum, Cambridge (1993), p. 30.
30 In Marie Lødrup Bang, *Johan Christian Dahl (1788–1857): Life and Works* (Oslo, 1987), 3 vols.
31 Clarkson Stanfield to his wife Rebecca, 16 January 1839, quoted Peter Van Der Merwe, *The Spectacular Career of Clarkson Stanfield 1793–1867*, exh. cat., Tyne and Wear Museums, Newcastle (1979), no. 218.
32 The most powerful eruptions in this period were Agung (Bali): 1808; Souffrière (West Indies): 1812; Mayon (Philippines): 1814; Tambora (Java): 1815; Raung (Java): 1817; Colima (Mexico): 1818; Usu (Japan): 1822; Galunggung (Java): 1822; Katla (Iceland): 1823; Grimsvotn (Iceland): 1823, 1838, 1854; Lanzaroti (Canary Islands): 1824; Isanotki (Aleutian Islands: 1825; Kelut (Java): 1826; Kliuchevskoi (Kamchatka): 1829; Cosiguina (Nicaragua): 1835; Ngauruhoe (New Zealand): 1839; Hekla (Iceland): 1845–6; Fonualei (Tonga): 1847; Cotopaxi (Ecuador): 1850–63; Chikurachki (Japan): 1853–9; Shiveluch (Kamchatka): 1854; Fuego (Guatemala): 1857.
33 Alexander von Humboldt, *Personal Narrative of Travels to the Equinocial Regions of the New Continent during the years 1799-1804* (London 1815), p. 384.
34 Alexander von Humboldt, *Cosmos: Sketch of a Physical Description of the Universe* (London, 1847–58), vol. II, p. 93.
35 David C. Huntington, 'Landscape and Diaries; The South American Trips of F. E. Church', *Brooklyn Museum Annual*, 5 (1963–4).
36 Andrew Wilton and Tim Barringer, *American Sublime: Landscape Painting in the United States 1820–1880* (London, 2002), pp. 220–21.
37 W. P. Bayley, 'Mr Church's Pictures', *Art-Journal* (September 1865), p. 265.
38 John K. Howat, *Frederic Church* (New Haven, CT, 2005), p. 47.
39 John Ruskin, *Modern Painters* (London, 1843–60), vol. IV, part V, chap. 7, para. 5.
40 John Brett, description of Taormina, Sicily, 17 January [1871], in *Nature*, III, 2 February 1871, p. 266.

41 Allen Staley and Christopher Newell, *Pre-Raphaelite Vision – Truth to Nature*, exh. cat., Tate, London (2004), no. 101.
42 Jules Verne, *Journey to the Centre of the Earth* [1864] (Oxford, 1992), pp. 209ff.
43 Erroll Sherson, *London's Lost Theatres of the Nineteenth Century* (London, 1992), pp. 204–5.

6 Krakatoa Shakes the World

1 Simon Winchester, *Krakatoa: The Day the World Exploded* (London, 2003).
2 Ibid., pp. 155ff.
3 University of Birmingham Research and Cultural Collections, P02442.
4 'Volcanic eruptions in Java', *The Times*, 30 August 1883.
5 Ferdinand de Jupilles, quoted in Jonathan Ribner, 'The Poetics of Pollution', *Turner Whistler Monet*, ed. Katharine Lochnan, exh. cat., Tate (London, 2004), p. 53. See also James Hamilton, *London Lights – The Minds that Moved the City that Shook the World* (London, 2007), chap. 1.
6 Gustave Kahn, on Monet's paintings of London, exhibited in 1904. Quoted Lochnan, ed., *Turner Whistler Monet*, cat. 74–7, p. 192.
7 William Ascroft, preface to *Catalogue of Sky Sketches from September 1883 to September 1886, illustrating optical phenomena attributed to the eruption of Krakatoa, in the Java Straits, August 27th, 1883*, Department of Science and Art (South Kensington Museum), 1883.
8 Ibid., p. 1.
9 Ibid., p. 4.
10 In 1884, 1885 and 1886.
11 Letter from Ascroft in the Files of the Krakatoa Committee, Royal Society Archive, quoted Thomas A. Zaniello, 'The Spectacular English Sunsets of the 1880s', *Annals of the New York Academy of Sciences*, CCCLX: *Victorian Science and Victorian Values: Literary Perspectives* (1981), pp. 247–67.
12 Gerard Hopkins [*sic*], 'The Remarkable Sunsets', *Nature*, XXIX (3 January 1884), pp. 222–3
13 R. M. Ballantyne, *Blown to Bits* (London, 1889), p. 374.
14 *New York Times*, 28 November 1883.
15 Edvard Munch, *Diary*, 22 January 1892; José Maria Faerna, *Munch* (New York, 1995), p. 17.
16 'Letter to the Editor: The Krakatoa Eruption', *The Times*, 13 February 1884.
17 G. J. Symons FRS, ed., 'Hon. Rollo Russell on the Unusual Twilight Glows', *The Eruption of Krakatoa, and Subsequent Phenomena*,

Report of the Krakatoa Committee of the Royal Society (1888), p. 159.
18 Ibid., pp. 384–405.
19 Log account of eruption of Krakatoa, Lapworth Museum, University of Birmingham, L161.

7 'The Night had Vanished': Vorticism and the Volcano

1 Entry for 8 May 1902, in *A Victorian Diarist – Later Extracts from the Journals of Mary, Lady Monkswell, 1895–1909*, ed. the Hon. E.F.C. Collier (London, 1946), p. 87.
2 'Letter to the Editor: The Great Eruption of 1868 in Hawaii', *The Times*, 17 May 1902.
3 Patrick Leigh Fermor, *The Violins of Saint-Jacques* [1953] (Harmondsworth, 1961), p. 10.
4 Ibid., p. 36.
5 Ibid., p. 108.
6 Ibid., p. 118.
7 Ibid., p. 125.
8 Ibid., p. 14.
9 Teresa A. Carbone, ed., *American Painters in the Brooklyn Museum: Artists born by 1876* (London and New York, 2006) vol. I, pp. 396–8.
10 The original reads: '*La cenere, l'oscurità, i boati, la paura percorrendo la via da Napoli a Resina* / E. Dalbono 11 Aprile 1906'.
11 For information on twentieth-century Icelandic art see Julian Freeman, ed., *Landscapes from a High Latitude: Icelandic Art 1909–1989* (London, 1989); and *Confronting Nature: Icelandic Art of the 20th Century*, exh. cat., Corcoran Gallery of Art, Washington, DC (2001).
12 I am grateful to Ari Trausti Gudmundsson and Orri Magnusson for information about Gudmundur Einarsson, Ari Trausi Gudmundsson's father.
13 'Atl' means 'water' in the indigenous Aztec language Nahuatl. Information on Dr Atl from the essay by Simon Grant in the booklet *The Fall: Stefan Brüggeman and Dr Atl*, produced by Bloomberg LP (2008).
14 Ibid., unpaginated.
15 Eugène Ionesco, in the Introduction to Katia and Maurice Krafft, *Volcano* (New York, 1975).

8 The Shifting Furnace

1 Now known to migrate from the Arctic to Europe in the spring, these birds seen flying from beyond the horizon were in the Middle Ages

believed to have come out of the sea, born from shellfish or driftwood.

2 Dirk Dobke, *Dieter Roth: Graphic Works Catalogue Raisonné, 1947–1998* (Stuttgart, London, Rekjavík, 2003), pp. 18 and 202–7.

3 John McEwan, *The Sculpture of Michael Sandle* (Aldershot, 2002), pp. 49–51.

4 Ilana Halperin, 'Geologic Intimacy', at www.ilanahalperin.com/new/statement.html, accessed 6 April 2011.

5 Ian Brown, communication with author.

6 Ilana Halperin, 'Geologic Intimacy', at www.ilanahalperin.com/new/statement.html, accessed 6 April 2011.

7 The Geological Society of London, *Super-eruptions: Global Effects and Future Threats* (London, 2005).

SELECT BIBLIOGRAPHY

Beard, Mary, *Pompeii: The Life of a Roman Town* (London, 2008)
Bullard, Fred M., *Volcanoes of the Earth* (St Lucia, Queensland, 1977)
Darley, Gillian, *Vesuvius: The Most Famous Volcano in the World* (London, 2011)
Edmaier, Bernhard, *Earth on Fire: How Volcanoes Shape our Planet* (London, 2009)
Geological Society of London, *Super-eruptions: Global Effects and Future Threats* (London, 2005)
Gudmundsson, Ari Trausti, *Living Earth: Outline of the Geology of Iceland* (Reykjavík, 2007)
Jenkins, Ian, and Kim Sloan, *Vases and Volcanoes: Sir William Hamilton and his Collections*, exh. cat., British Museum (London, 1996)
Krafft, Maurice and Katya, *Volcanoes: Fire from the Earth* (London, 1993)
Muséum Genève, *Supervolcan* (Geneva, 2011)
Rothery, David, *Volcanoes, Earthquakes and Tsunamis* (London, 2001)
Scarth, Alwyn, *Volcanoes, An Introduction* (London, 1994)
——, and Jean-Claude Tanguy, *Volcanoes of Europe* (Oxford, 2001)
Siebert, Lee, Tom Simkin, and Paul Kimberley, *Volcanoes of the World* (Washington, DC, 1995)
Sigurdsson, Haraldur, ed., *Encyclopedia of Volcanoes* (London, 1999)
Winchester, Simon, *Krakatoa: The Day the World Exploded* (New York, 2003)

Introductions for Younger Readers

Greenwood, Rosie, *I Wonder why Volcanoes Blow their Tops, and Other Questions about Natural Disasters* (Kingfisher, 2004)
Schreiber, Anne, *Volcanoes!* (Washington, DC, 2007)
Turnbull, Stephanie, *Volcanoes* (London, 2007)

Fiction

Bulwer Lytton, Edward, *The Last Days of Pompeii* (1834)
Harris, Robert, *Pompeii* (New York, 2003)
Leigh Fermor, Patrick, *The Violins of Saint-Jacques* (London, 1953)
Lowry, Malcolm, *Under the Volcano* (London, 1947)
Rhys, Jean, 'Heat', in *The Collected Short Stories of Jean Rhys* (New York, 1987)
Sigurdardottir, Yrsa, *Ashes to Dust* (London, 2010)
Sontag, Susan, *The Volcano Lover – A Romance* (London, 1992)
Verne, Jules, *Journey to the Centre of the Earth* (1864)

ASSOCIATIONS AND WEBSITES

The website of the Global Volcanism Program, run from the Smithsonian Institution in Washington, DC, presents what is perhaps the fullest list of major volcanic eruptions of magnitude Volcanic Explosivity Index (VEI) 4 or more, from all over the world. Going back only the relatively modest length of time, about 12,000 years, these make a total of 2103 events, from 9650BC. The site carries, further, encyclopaedic information on all aspects of volcanoes, active, quiescent and extinct.

www.volcano.si.edu

An independent volcano website in the UK is the Volcanism Blog, managed by Dr Ralph Harrison, while another is the enthusiastic, youthful and informative blog managed by Augusta Ward,

www.volcanism.wordpress.com
www.volcano-club.blogspot.com

The British Geological Survey publishes on its website a fact sheet entitled 'Why No Volcanoes?', explaining clearly why there are no active volcanoes in Britain.

www.bgs.ac.uk/home.html

A useful world volcanic map is at

www.volcano.si.edu/world/find_regions.cfm

This site carries further encyclopaedic information, including a click-on global map on all aspects of volcanoes, active, quiescent or extinct.

The French-language journal *Eruptions*, edited by Frédéric Lécuyer, carries current news on volcano matters, and articles about volcanic activity worldwide, both contemporary and historic.

www.pyros-volcans.com

The Krafft Collection of pictures is held at the Bibliothèque centrale du Muséum national d'Histoire naturelle, 38, rue Geoffroy Saint-Hilaire, 75005 Paris.

www.mnhn.fr/expo/expo10ans/krafft001.htm
www.imagesdevolcans.fr

In memory of the volcanologists Maurice and Katya Krafft, who were killed by a pyroclastic flow on Mount Unzen, Japan, in 1991, the Krafft Medal is awarded every four years by the Krafft family and the International Association of Volcanology and Chemistry of the Earth's Interior (IAVCEI). The medal is given to those who have shown 'altruism and dedication to the humanitarian and applied sides of volcanology and those who have made selfless contributions to the volcanological community.' Past winners have been Tom Simkin (2004) and Christopher G. Newhall (2008).

www.iavcei.org/IAVCEI.htm

Other websites include

www.volcan.ch (Société de Volcanologie, Geneva)

www.swisseduc.ch/stromboli

www.decadevolcano.net

www.nps.gov/havo

ACKNOWLEDGEMENTS

This book erupted as a consequence of the exhibition 'Volcano: From Turner to Warhol', held at Compton Verney, Warwickshire, in the summer of 2010. It was the first exhibition in Britain to explore and celebrate the extraordinary range of artists' responses to volcanoes, and their change over time. The show was beautiful, exciting and popular, and drew lively attention to the fertility, variety and longevity of its subject.

The initial idea of the show was suggested to me in 2007 by Kathleen Soriano when she was Director of Compton Verney. Her encouragement, enthusiasm, and her invitation to me to be the guest curator was echoed and maintained by her successor as Director, Dr Stephen Parissien, and his many colleagues, including Christine Cluley, Alison Cox, Jack Crossley, Verity Elson, David Jones, John Leslie, Jenine McGaughran, Abi Pole and Amanda Randall. I am most grateful to Steven Parissien for allowing me to publish this enlarged version of the text in the exhibition's accompanying booklet. This was dedicated to the memory of the volcanologists Maurice and Katya Krafft.

Many generous friends and colleagues aided the exhibition and this publication with advice and guidance. Among them in Britain, France, Iceland, Italy and the USA are Dr Patricia Andrew, Eleanor Antin, Huginn Aranson, Professor Renato Barisani, Emilie Beck-Saiello, Bettina Bélanger, Dr Dominique Bertrand, De Soto Brown, Ian Brown, Lucio Capelli, David Clarkson, Giovanni and Anna Rosa Cotroneo, Brigitte Dapra, James P. Graham, Ari Trausti Gudmundsson, Ilana Halperin, Colin Harris, Jane Harrison, Gill Hedley, Dagny Heidal, Professor Frank James, Olafur Ingi Jonsson, James Miller, Rossana Muzii, Dr Clare Pollard, Bjorn Roth, Halldor Bjorn Runolfsson, Professor Renato Ruotolo, Michael Sandle, Dr Françoise Serre, Lee Shephard, Karen Sherry, Professor Aurora Spinosa, Professor Nicola Spinosa, and HafthorYngvason. At the University of Birmingham my colleagues Dr Jon Clatworthy, Professor Ken Dowden, Clare Mullett, Professor Paul Smith, Dr Gillian Shepherd and Dr Carl Stevenson were constant sources of encouragement. Particular thanks for help and guidance go also to Felicity Bryan, Professor Oswyn Murray, and, as always, to my family.

I am indebted also to Daniel Allen and Michael Leaman of Reaktion Books, and, for permissions to quote from copyright material, to Eleanor Antin, the Bodleian Library (University of Oxford), the British Library Board, Ian Brown, Ilana Halperin, the Institution of Engineering and Technology, the Lapworth Museum and the Research and Cultural Collections, University of Birmingham, John R. Murray (for Patrick Leigh Fermor), the North Yorkshire County Record Office, Penguin Books (for Susan Sontag), the Royal Institution of Great Britain, the Royal Society, and the Virginia Historical Society. Every effort has been made to contact copyright holders, and any omission is inadvertent. Omissions will be rectified in future editions if copyright holders will contact me.

PHOTO ACKNOWLEDGEMENTS

The author and the publishers wish to express their thanks to the below sources of illustrative material and/or permission to reproduce it:

© The Andy Warhol Foundation for the Visual Arts / Artists Rights Society (ARS), New York / DACS, London 2012: p. 178; Art Museum of Bergen, Norway: p. 119; Work in the collection of the British Museum, London, image courtesy of Art First, London: p. 175; Ashmolean Museum of Art and Archaeology, Oxford: p. 34 (Jennings / Spaldings Gift, EAX. 4381); Courtesy of Norinne Betjemann: p. 164; Bibliothèque Centrale du Museum National d'Histoire Naturelle, Paris: pp. 35 (Krafft Gift MNHN.OA.KR.9), 42 (Krafft Gift MNHN.OA.KR.50), 63 (Krafft Gift MNHN.OA.KR.57), 77 (Krafft Gift MNHN.OA.KR.24), 98 bottom (MNHN.OA.KR.28); The British Library, London: p. 66; © The Trustees of the British Museum, London: pp. 12, 19, 60, 65, 100, 101, 102, 126; Courtesy of Ian Brown: p. 172; Corbis: pp. 117 (Brooklyn Museums), 148 (Brooklyn Museums); Courtesy of David Clarkson: p. 37; Compton Verney: pp. 70 (0357.B), 79 (0343.S); © DACS 2012: p. 157; Derby Museum and Art Gallery, Derby: pp. 78, 83; Courtesy of the artist and Dundee Contemporary Arts Print Studio: p. 169; Courtesy Ronald Feldman Fine Arts, New York: p. 173; Fitzwilliam Museum, University of Cambridge: p. 168; Gallery of the Academy of Fine Arts, Naples: pp. 149 (inv. no. 382); Courtesy of James P. Graham: pp. 170–71; Wilhelmina Barns-Graham Trust: p. 38; Courtesy of Ilana Halperin: p. 39; Istockphoto: pp. 6 (Rainer Albiez), 21 (Josef Friedhuber), 43 (Josef Friedhuber); Library of Congress, Washington, DC: pp.132, 142; © Munch Museum / Munch –Ellingsen Group, BONO, Oslo / DACS, London 2012: p. 138; John Murray Publishers: p. 145; National Gallery of Iceland, Reykavik / Myndstef – Iceland Visual Art Copyright Association: pp. 34, 98, 99 (top), 151; National Gallery of Norway, Oslo: p. 158; © National Maritime Museum, Greenwich, London: p. 106; Paul K: pp. 72, 73, 75; Private Collection, on permanent loan to the

Gallery of the Academy of Fine Arts, Naples: p. 159; Rex Features: pp. 23 (Fotos International), 23–24 (Martin Rietze / WestEnd61), 28 (Denis Cameron), 31 (WestEnd61), 33 (Tony Waltham / Robert Harding), 40-41 (Moodboard), 177 (Sipa Press), 179 (East News), 181 (Alfo); Reykajavik Art Musem / Mydsef – Icelandic Visual Arts Copyrights Association: p. 152; The Royal Society, London: p. 71; © Dieter Roth Estate, Courtesy of Hauser & Wirth: p. 163; Courtesy of Michael Sandle: p. 166; Gerald Scarfe: p. 103; Science and Society Picture Library: p. 135; Sheffield Galleries and Museums Trust: p. 125; Southampton City Art Gallery: p. 96; The State Russian Museum, St Petersburg: p. 111; Statens Museum for Kunst, Copenhagen: p. 119 (inv. no. 6803); © Tate, London 2012: pp. 80, 81, 92, 95, 120; Courtesy of Ari Trausti: p. 154; Yale Center for British Art, Paul Mellon Collection, New Haven, Connecticut: p. 93.

INDEX